Field-Programmable Gate Arrays

AF654579

Field-Programmable Gate Arrays

Stephen D. Brown
University of Toronto

Robert J. Francis
University of Toronto

Jonathan Rose
University of Toronto

Zvonko G. Vranesic
University of Toronto

Springer

Library of Congress Cataloging-in-Publication Data

Field-programmable gate arrays / Stephen D. Brown... [et al.].
cm. -(Kluwer international series in engineering
and computer science ; SECS 180)
Includes bibliographical references and index.
ISBN 0-7923-9248-5
1. Programmable logic devices. 2. Gate array circuits.
I. Brown, Stephen D. II. Series
TK7872.L64F54 1994
621.39'5—dc20 92-13785
CIP

All rights reserved. No part of this publication may be reproduced, stored in a retrieval system or transmitted in any form or by any means, mechanical, photo-copying, recording, or otherwise, without the prior written permission of the publisher.

Brown, Francis, Rose & Vranesic: Field-Programmable Gate Arrays

© 1992 Kluwer Academic Publishers
(Now a part of Springer Science+Business Media)

All rights reserved. No part of this publication may be reproduced, stored in any electronic or mechanical form, including photocopy, recording or otherwise, without the prior written permission of the publisher.

First Indian Reprint 2007
Third Indian Reprint 2014

ISBN 978-81-8128-689-5

"This edition is licensed for sale only in India, Pakistan, Bangladesh, Sri Lanka and Nepal. Circulation of this edition outside of these territories is UNAUTHORIZED AND STRICTLY PROHIBITED."

This edition is published by Springer (India) Private Limited, (a part of Springer Science+Business Media), Registered Office: 7th Floor, Vijaya Building, 17, Barakhamba Road, New Delhi - 110 001, India

Printed in India by Rashtriya Printers, Delhi.

To Susan,
Ming,
Barbara, Jessica, Hannah,
and Anne

Contents

Preface

This book deals with Field-Programmable Gate Arrays (FPGAs), which have emerged as an attractive means of implementing logic circuits, providing *instant* manufacturing turnaround and negligible prototype costs. They hold the promise of replacing much of the VLSI market now held by Mask-Programmed Gate Arrays. FPGAs offer an affordable solution for customized VLSI, over a wide variety of applications and have also opened up new possibilities in designing reconfigurable digital systems.

The book discusses the most important aspects of FPGAs in a textbook manner. It is not an edited collection of papers. It gives the reader a focused view of the key issues, using a consistent notation and style of presentation. It provides detailed descriptions of commercially available FPGAs and an in-depth treatment of the FPGA architecture and CAD issues that are the subjects of current research.

The material presented will be of interest to a variety of readers. In particular, it should appeal to:

1. Readers who are not familiar with FPGA technology, but wish to be introduced to it. They will find an extensive survey that includes products from ten FPGA manufacturers, and a discussion of the most pertinent issues in the design of FPGA architectures, as well as the CAD tools needed to make effective use of them.

2. Readers who already have an understanding of FPGAs, but who are interested in learning about the research directions that are of current interest.

Chapter 1 introduces FPGA technology. It defines an FPGA to be a user-programmable integrated circuit, consisting of a set of logic blocks that can be interconnected by general routing resources. A survey of commercial FPGA devices is provided in Chapter 2. This includes descriptions of the chip architectures and the basic technologies that are needed to to achieve the programmability. Chapter 3 deals with the Computer-Aided Design (CAD) task known as "technology mapping," which determines how a given logic circuit can be implemented using the logic blocks available in a particular FPGA. Included are examples of technology mapping algorithms for two types of FPGA. Chapter 4 considers the design of the logic block and its effect on the speed and logic density of FPGA circuits. It gives the results of several recent studies on this topic. The next chapter focuses on the CAD routing problem in FPGAs, where the interconnections between the logic blocks are realized. Examples of algorithms are presented for two different types of FPGA. Chapter 6 investigates the question of how the richness of the routing resources affects the FPGA's ability to implement circuits. It shows the results of a recent experimental study. The final Chapter also considers the the routing resources, but uses a mathematical modelling technique. This provides an example of how FPGAs can be studied and improved through theoretical research.

The authors wish to acknowledge the encouragement and help of Carl Harris, of Kluwer Academic Publishers, who has ensured that this book was produced in optimum time. We would also like to express our appreciation to the many members of the FPGA research project at the University of Toronto, whose efforts have contributed both to the information presented in this book and to the general understanding of the many complex issues in the design and use of FPGAs. These include Professors Paul Chow and David Lewis, as well as Kevin Chung, Bahram Fallah, Keith Farkas, Alan Huang, Carl Mizuyabu, Gerard Paez, Immanuel Rahardja, Soon Ong Seo, Satwant Singh, Benjamin Tseng, and Jean-Michel Vuillamy. Professor Mart Molle provided valuable comments on the stochastic modelling chapter. Jack Kouloheris and Abbas El Gamal of Stanford University generously provided several figures and engaging discussions. The authors gratefully acknowledge enlightening conversations with many people in the FPGA industry and environs. In particular Steve Trimberger, Bill Carter and Erich Goetting from Xilinx, Jonathan Greene and Andy Haines at Actel, Stan Kopec and Clive McCarthy from Altera, Dwight Hill from AT&T Bell Labs, and David Marple from Crosspoint.

Glossary

Anti-Fuse

a programming element switch which is normally open, and which closes when a high voltage is placed across its terminals.

Area-efficiency (of an FPGA architecture)

the amount of area required by the architecture to implement a given amount of logic circuitry.

Binary Decision Diagram (BDD)

a method of representing Boolean logic expressions using a selector element and Shannon decomposition..

Channel

the rectangular area that lies between two rows or two columns of logic blocks. A routing channel contains a number of tracks.

Channel Density

the maximum number of connections in parallel anywhere in a channel.

Channel Segment
a section of the routing channel.

Connection Block
a structure in the routing architecture of an FPGA that provides connections between the pins of the logic block and the routing channels.

EEPROM
Electrically Erasable Programmable Read Only Memory.

EPROM
Erasable Programmable Read Only Memory.

Field-Programmable Device
a device that can be configured by the user with simple electrical equipment.

Flexibility (of routing architecture)
the number of choices offered by a routing architecture in making a set of connections.

FPGA Architecture
the logic block, routing and I/O block structure of an FPGA.

F_c a parameter specifying connection block flexibility.

F_s a parameter specifying switch block flexibility.

Global Router
a CAD tool that determines which set of channels each connection travels through.

Logic Block
the basic unit of the FPGA that performs the combinational and sequential logic functions.

Logic Block Architecture
the choice of combinational and sequential functions of the logic block,

and their interconnection within that block.

Logic Block Functionality
the number of different combinational functions that a logic block can implement.

Logic Density (of an FPGA)
the amount of logic capability per unit area that an FPGA achieves.

Lookup Table (LUT)
a digital memory with K address lines that can implement any function of K inputs by placing the truth table into the memory.

Mask-Programed Gate Array (MPGA)
an IC with uncommitted arrays of transistors that are personalized by two or more layers of metal connections.

PAL Programmable Array Logic.

Pass Transistor
a transistor used as a switch to make a connection between two points.

Placement
the CAD task of assignment of logic blocks to physical locations.

PLD Programmable Logic Device.

Programmable Inversion
a feature of a logic block which allows that inputs or outputs can be programmed in true or complemented form.

Programming Technology
the fundamental method of customization in an FPGA that provides the user-programmability. Examples are SRAM, anti-fuse, EPROM and EEPROM.

Programmable Switch
a switch in an FPGA that is used to connect two wire segments, and can

be programmably opened or closed using the programming technology.

Routability
the percentage of required connections successfully completed after routing.

Routing Architecture
the distribution and length of wire segments, and the manner in which the wire segments and programmable switches are placed in the routing channels.

Segmented Channel
a routing channel where tracks contain wire segments of varying lengths.

Switch Block
a structure in the routing architecture which connects one routing channel to another.

Technology Mapping
the CAD task of converting boolean expressions into a network that consists of only logic blocks.

Track (routing)
a straight section of wire that spans the entire width or length of a routing channel. A track can be composed of a number of wire segments of various lengths.

Wire Segment
a length of metal wire that has programmable switches on either end, and possibly switches connected to the middle of the wire. It cannot be broken by a programmable switch, or else it would be two wire segments.

Field-Programmable Gate Arrays

CHAPTER 1
Introduction to FPGAs

Very Large Scale Integration (VLSI) technology has opened the door to the implementation of powerful digital circuits at low cost. It has become possible to build chips with more than a million transistors, as exemplified by state-of-the-art microprocessors. Such chips are realized using the *full-custom* approach, where all parts of a VLSI circuit are carefully tailored to meet a set of specific requirements. Semi-custom approaches such as Standard Cells and Mask-Programmed Gate Arrays (MPGAs) have provided an easier way of designing and manufacturing Application-Specific Integrated Circuits (ASICs).

Each of these techniques, however, requires extensive manufacturing effort, taking several months from beginning to end. This results in a high cost for each unit unless large volumes are produced, because the overhead to begin production of such chips ranges from $20,000 to $200,000.

In the electronics industry it is vital to reach the market with new products in the shortest possible time, and so reduced development and production time is essential. Furthermore, it is important that the financial risk incurred in the development of a new product be limited so that more new ideas can be prototyped. Field-Programmable Gate Arrays (FPGAs) have emerged as the ultimate solution to these time-to-market and risk problems because they provide instant manufacturing and very low-cost prototypes. An FPGA can be manufactured in only minutes, and prototype costs are on the order of $100. A field-programmable device is a device in which the final logic structure can be directly configured by the end user, without the use of an integrated circuit fabrication facility.

The last three years have seen FPGAs grow from a tiny market niche into a $200 million business. It is expected that almost one billion dollars worth of FPGAs will be sold every year by 1996, representing a significant proportion of the IC market.

This book is concerned with many aspects of FPGA architecture and the Computer-Aided Design Tools needed in their use. This chapter begins by describing the evolution of programmable devices and gives a brief introduction to FPGAs, their economics and their use. It also provides an indication of the material presented in subsequent chapters.

1.1 Evolution of Programmable Devices

Programmable devices have long played a key role in the design of digital hardware. They are general-purpose chips that can be configured for a wide variety of applications. The first type of programmable device to achieve widespread use was the *Programmable Read-Only Memory* (PROM). A PROM is a one-time programmable device that consists of an array of read-only cells. A logic circuit can be implemented by using the PROM's address lines as the circuit's inputs, and the circuit's outputs are then defined by the stored bits. With this strategy, any truth-table function can be implemented.

Two basic versions of PROMs are available, those that can be programmed only by the manufacturer, and those that can be programmed by the end-user. The first type is called *mask-programmable* and the second is *field-programmable.* In the context of implementing logic circuits, superior speed-performance can be obtained with a mask-programmable chip because connections within the device can be hardwired during manufacture. In contrast, field-programmable connections always involve some sort of programmable *switch* (such as a fuse) that is inherently slower than a hardwired connection. However, a field-programmable device offers advantages that often outweigh its speed-performance shortcomings:

- Field-programmable chips are less expensive at low volumes than mask-programmable devices because they are standard off-the-shelf parts. An IC manufacturing facility must be "tooled" to begin production of a mask-programmed device which incurs a large overhead cost.
- Field-programmable chips can be programmed immediately, in minutes, whereas mask-programmable devices must be manufactured by a foundry over a period of weeks or months.

Two field-programmable variants of the PROM, the *Erasable Programmable Read-Only Memory* (EPROM) and the *Electrically Erasable Programmable Read-Only Memory* (EEPROM) offer an additional advantage; both can be erased and re-programmed many times. In some applications, and particularly during the early stages of a logic circuit's design, re-programmability is an attractive feature.

While PROMs are a viable alternative for realizing simple logic circuits, it is clear that the structure of a PROM is best suited for the implementation of computer memories. Another type of programmable device, designed specifically for implementing logic circuits, is the *Programmable Logic Device* (PLD). A PLD typically comprises an array of AND gates connected to an array of OR gates. A logic circuit to be implemented in a PLD is thus represented in sum-of-products form. The most basic version of a PLD is the *Programmable Array Logic* (PAL). A PAL consists of a programmable AND-plane followed by a fixed OR-plane. The outputs of the OR gates can be optionally registered by a flip-flop in most chips. PALs also offer the advantages of field-programmability, which is obtained using one of fuse, EPROM or EEPROM technology.

A more flexible version of the PAL is the *Programmable Logic Array* (PLA). PLAs also comprise an AND-plane followed by an OR-plane, but in this case connections to both planes are programmable. They are available in both mask-programmable and field-programmable versions.

With their simple two-level structure, both types of PLDs described above allow high speed-performance implementations of logic circuits. However, the simple structure also leads to their main drawback. They can only implement small logic circuits that can be represented with a modest number of product terms, because their interconnection structure would grow impractically large if the number of product terms were increased.

The most general type of programmable devices consists of an array of uncommitted elements that can be interconnected according to a user's specifications. Such is the class of devices known as *Mask-Programmable Gate Arrays* (MPGAs). The most popular MPGAs consist of rows of transistors that can be interconnected to implement a desired logic circuit. User-specified connections are available both within the rows (to implement basic logic gates) and between the rows (to connect the basic gates together). In addition to the rows of transistors, some circuitry is provided that handles input and output to the external pins of the IC package. In an MPGA, all the mask layers that define the circuitry of the chip are pre-defined by the manufacturer, except those that specify the final metal layers. These metal layers are customized to connect the transistors in the array, thereby implementing the desired circuit. MPGAs have a large non-recurring engineering (NRE) cost because of the need to generate the metal mask layer and manufacture the chip. However, the unit cost decreases significantly when large volumes (more than 1000 chips) are required.

The main advantage of MPGAs over PLDs is that they provide a general structure that allows the implementation of much larger circuits. This is primarily due to their interconnection structure, which scales proportionally with the amount of logic. On the other hand, since MPGAs are mask-programmable, they require significant manufacturing time and incur high initial costs. A Field-Programmable Gate Array combines the programmability of a PLD and the scalable interconnection structure of an MPGA. This results in programmable devices with much higher logic density.

1.2 What is an FPGA?

Like an MPGA, an FPGA consists of an array of uncommitted elements that can be interconnected in a general way. Like a PAL, the interconnections between the elements are user-programmable. FPGAs were introduced in 1985 by the Xilinx Company. Since then, many different FPGAs have been developed by a number of companies: Actel, Altera, Plessey, Plus, Advanced Micro Devices (AMD), QuickLogic, Algotronix, Concurrent Logic, and Crosspoint Solutions, among others. Chapter 2 describes the FPGAs produced by each of these ten companies.

Figure 1.1 shows a conceptual diagram of a typical FPGA. As depicted, it consists of a two-dimensional array of logic blocks that can be connected by general interconnection resources. The *interconnect* comprises segments of wire, where the segments may be of various lengths. Present in the interconnect are programmable switches that serve to connect the logic blocks to the wire segments, or one wire segment to another. Logic circuits are implemented in the FPGA by partitioning the logic into individual logic blocks and then interconnecting the blocks as required via the switches.

To facilitate the implementation of a wide variety of circuits, it is important that an FPGA be as versatile as possible. This means that the design of the logic blocks, coupled with that of the interconnection resources, should facilitate the implementation of a large number of digital logic circuits. There are many ways to design an FPGA, involving tradeoffs in the complexity and flexibility of both the logic blocks and the interconnection resources. This book will address most of the relevant issues involved.

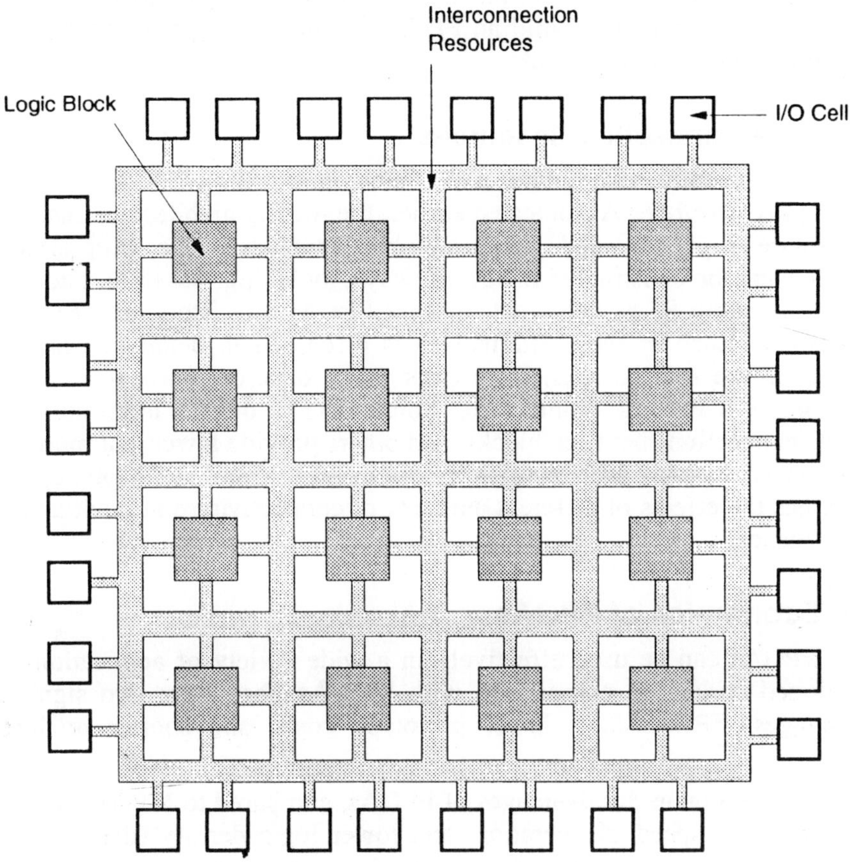

Figure 1.1 - *A Conceptual FPGA.*

1.2.1 Logic Blocks

The structure and content of a logic block are called its *architecture*. Logic block architectures can be designed in many different ways. As shown by the examples in Chapter 2, some FPGA logic blocks are as simple as 2-input NAND gates. Other blocks have more complex structure, such as multiplexers or lookup tables. In some FPGAs, a logic block corresponds to an entire PAL-like structure. There exists a myriad of possibilities for defining the logic block as a more complex circuit, consisting of several sub-circuits and having more than one output. Most logic blocks also contain some type of flip-flop, to aid in the implementation of sequential circuits. Logic block architecture is discussed in detail in Chapter 4. Included are the results of studies that show the effects of the architecture of the logic block on both the total chip area needed to build an FPGA and the speed performance of circuits implemented in an FPGA.

1.2.2 Interconnection Resources

The structure and content of the interconnect in an FPGA is called its *routing architecture*. As indicated earlier, the routing architecture consists of both wire segments and programmable switches. The programmable switches can be constructed in several ways, including: pass-transistors controlled by static RAM cells, anti-fuses, EPROM transistors, and EEPROM transistors. Each of these alternatives is discussed in detail in Chapter 2. Similar to the logic blocks, there exist many different ways to design the structure of a routing architecture. Some FPGAs offer a large number of simple connections between blocks, and others provide fewer, but more complex routes. Routing architectures are discussed in detail in Chapter 6, which examines the effects of different amounts of connectivity on circuit area and performance.

1.3 Economics of FPGAs

FPGAs can be used effectively in a wide variety of applications. As mentioned earlier, in comparison with MPGAs, they have two significant advantages: FPGAs have lower prototype costs, and shorter production times.

The two main disadvantages of FPGAs, compared to MPGAs, are their relatively low speed of operation, and lower logic density (the amount of logic that can be implemeted in a single chip). The propagation delay of an FPGA is adversely affected by the inclusion of programmable switches, which have significant resistance and capacitance, in the connections between logic blocks. A direct comparison with MPGAs indicates that a typical circuit will be slower by a factor of roughly three if implemented in an FPGA.

Logic density is decreased because the programmable switches and associated programming circuitry require a great deal of chip area compared to the metal connections in an MPGA. Typical FPGAs are a factor of 8 to 12 times less dense than MPGAs manufactured in the same IC fabrication process. The larger area required by the FPGA for the same amount of logic circuitry means that fewer FPGA chips can be produced per wafer than in the case of an MPGA, and a lower yield is likely. At higher production volumes this means that an FPGA is much more expensive than an MPGA.

For example, consider a 2,000-gate MPGA and a 2,000-gate FPGA fabricated in the same IC process. In 1990, the manufacturing overhead cost of this MPGA was roughly $20,000, after which the cost of producing each individual chip was about $5. For the 2000-gate FPGA, there is no overhead cost, but the per chip cost is roughly $50, which may decrease to about $35 at higher volumes. Figure 1.2 illustrates these figures. At low volumes the FPGA unit cost is much lower than for an MPGA. As the figure indicates, the break-even point for the two technologies occurs at a volume of about 700. Note that this analysis does not include the cost of testing, inventory, or other costs that may affect an economic decision.

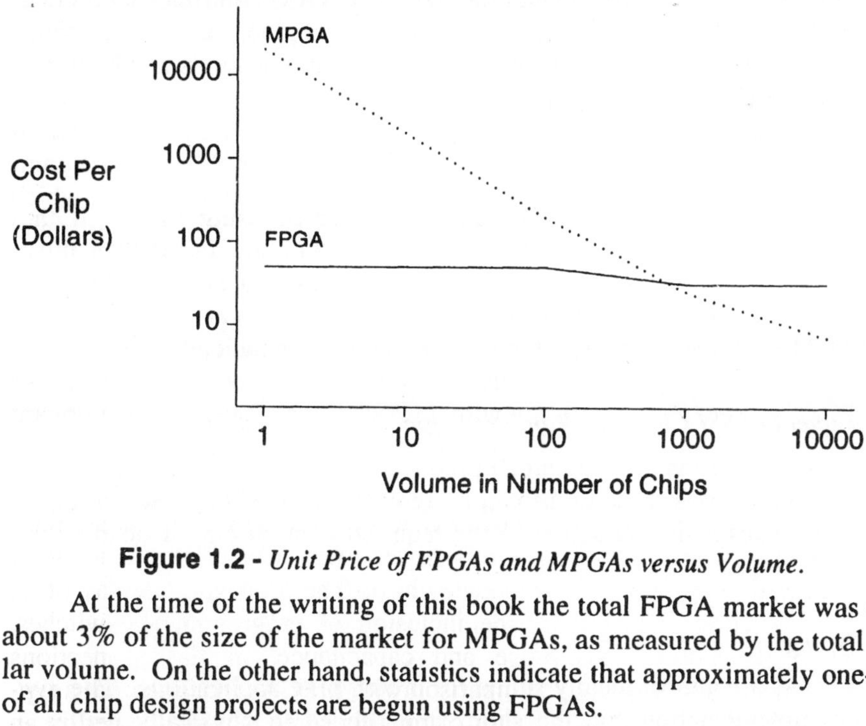

Figure 1.2 - *Unit Price of FPGAs and MPGAs versus Volume.*

At the time of the writing of this book the total FPGA market was only about 3% of the size of the market for MPGAs, as measured by the total dollar volume. On the other hand, statistics indicate that approximately one-half of all chip design projects are begun using FPGAs.

Several factors may contribute, in the future, to the emergence of the FPGA as a superior choice of implementation medium over the MPGA. For low gate counts, the difference in area between an FPGA and MPGA may be insignificant, because the chip size will be determined by the number of I/O pads and not the logic and interconnection. At this point the fabrication costs will be dominated by the package costs, which is the same for both technologies.

1.4 Applications of FPGAs

FPGAs can be used in almost all of the applications that currently use Mask-Programmed Gate Arrays, PLDs and small scale integration (SSI) logic chips. Below we present a few categories of such designs.

Application-Specific Integrated Circuits (ASICs)

An FPGA is a completely general medium for implementing digital logic. They are particularly suited for implementation of ASICs. Some examples of such use that have been reported are: a 1 megabit FIFO controller, an IBM PS/2 micro channel interface, a DRAM controller with error correction, a printer controller, a graphics engine, a T1 network transmitter/receiver as well as many other telecommunications applications, and an optical character recognition circuit.

Implementation of Random Logic

Random logic circuitry is usually implemented using PALs. If the speed of the circuit is not of critical concern (PALs are faster than most FPGAs), such circuitry can be implemented advantageously with FPGAs. Currently, one FPGA can implement a circuit that might require ten to twenty PALs. In the future, this factor will increase dramatically.

Replacement of SSI Chips for Random Logic

Existing circuits in commercial products often include a number of SSI chips. In many cases these chips can be replaced with FPGAs, which often results in a substantial reduction in the required area on circuit boards that carry such chips.

Prototyping

FPGAs are almost ideally suited for prototyping applications. The low cost of implementation and the short time needed to physically realize a given design, provide enormous advantages over more traditional approaches for building prototype hardware. Initial versions of prototypes can be implemented quickly and subsequent changes in the prototype can be done easily and inexpensively.

FPGA-Based Compute Engines

A whole new class of computers has been made possible with the advent of in-circuit re-programmable FPGAs. These machines consist of a board of such FPGAs, usually with the pins of neighboring chips connected. The idea is that a software program can be "compiled" (using high-level, logic-level and layout-level synthesis techniques, or by hand) into hardware rather than software. This hardware is then implemented by programming the board of FPGAs. This approach has two major advantages: first, there is no instruction fetching as required by traditional microprocessors, as the hardware directly embodies the instructions. This can result in speedups of the order of 100. Secondly, this computing medium can provide high levels of parallelism, resulting in a further speed increase.

The Quickturn company provides such a product tuned towards the simulation emulation of digital circuits. Also, Algotronix Ltd. sells a small add-in board for IBM PCs that can perform this function. At the research level, the Digital Equipment Corporation in Paris [Bert92] has achieved performance ranging from 25 billion operations per second up to 264 billion operations per second on applications such as RSA cryptography, the discrete cosine transform, Ziv-Lempel encoding and 2-D convolution, among others.

On-Site Re-configuration of Hardware

FPGAs are also attractive when it is desirable to be able to change the structure of a given machine that is already in operation. One example is computer equipment in a remote location that may have to be altered on site in order to correct a failure or perhaps a design error. A board that features a number of FPGAs connected via a programmable interconnection network allows a high degree of flexibility in augmenting the functional behavior of the circuitry provided by the board. Note that the most suitable type of FPGA for this kind of application is one that contains re-programmable switches.

1.5 Implementation Process

A designer who wants to make good use of FPGAs must have access to an efficient CAD system. Figure 1.3 shows the steps involved in a typical CAD system for implementing a circuit in an FPGA. Note that the system that is appropriate for each FPGA varies, and the one shown in the figure is only suggestive. An example of a real CAD system used for commercial FPGAs is presented in Chapter 2.

The starting point for the design process is the initial logic entry of the circuit that is to be implemented. This step typically involves drawing a schematic using a schematic capture program, entering a VHDL description, or specifying Boolean expressions. Regardless of the initial design entry, the circuit description is usually translated into a standard form such as Boolean

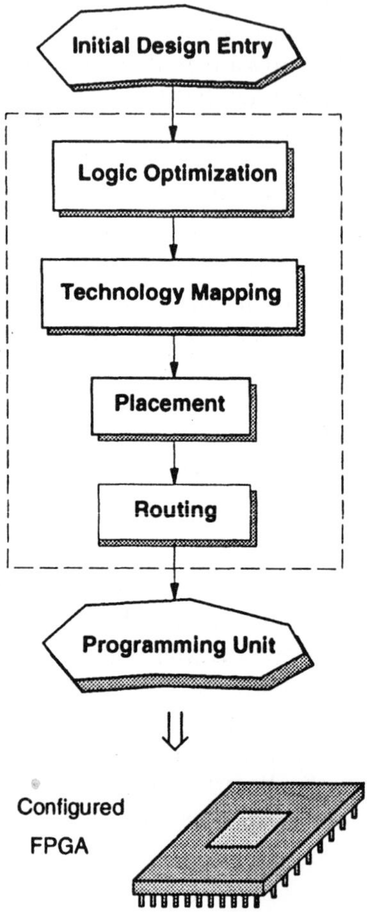

Figure 1.3 - *A Typical CAD System for FPGAs.*

expressions. The Boolean expressions are then processed by a logic optimization tool [Bray86, Greg86], which manipulates the expressions. The goal is to modify these expressions to optimize the area or speed of the final circuit. A combination of both area and delay requirements may also be considered. This optimization usually performs the equivalent of an algebraic minimization of the Boolean expressions and it is appropriate when implementing a logic circuit in any medium, not just FPGAs.

The optimized Boolean expressions must next be transformed into a circuit of FPGA logic blocks. This is done by a *technology mapping* program. The mapper may attempt to minimize the total number of blocks required, which is known as area optimization. Alternatively, the objective may be to minimize the number of stages of logic blocks in time-critical paths, which is called delay optimization. Technology mapping issues are dealt with in detail in Chapter 3, by presenting two examples of technology mapping algorithms for FPGAs.

Having mapped the circuit into logic blocks, it is necessary to decide where to place each block in the FPGA's array. A *placement* program is used to solve this problem. Typical placement algorithms attempt to minimize the total length of interconnect required for the resulting placement [Hana72, Sech87]. It should be noted that the problem of placement in the FPGA environment is quite similar to that in the case of VLSI circuits implemented with standard cells.

The final step in the CAD system is performed by the routing software, which assigns the FPGA's wire segments and chooses programmable switches to establish the required connections among the logic blocks. The routing software must ensure that 100 percent of the required connections are formed, otherwise the circuit cannot be realized in a single FPGA. Moreover, it is often necessary to do the routing such that propagation delays in time-critical connections are minimized. Routing in the FPGA environment involves similar concepts as in the standard cell environment, but it is complicated by the constraint that in FPGAs all of the available routing resources (wire segments and switches) are fixed in place. The routing issues for FPGAs are discussed in detail in Chapter 5, by presenting two examples of FPGA-specific routing algorithms.

Upon successful completion of the placement and routing steps, the CAD system's output is fed to a programming unit, which configures the final FPGA chip. The entire process of implementing a circuit in an FPGA can take from a few minutes to about an hour, depending on which FPGA is being used.

1.6 Concluding Remarks

This chapter has provided a brief introduction to FPGA technology. The chapters that follow will examine the most important issues at length. The intent of this material is not to give a "user manual" type of description, but to focus instead on the challenging questions pertinent to the design of FPGAs and the CAD tools needed to implement a user's logic circuit.

CHAPTER 2
Commercially Available FPGAs

Over the last few years, several companies have introduced a number of different types of FPGAs. While each product has unique features, they can all be classified into one of four categories. Figure 2.1 depicts the four main classes of FPGAs: symmetrical array, row-based, hierarchical PLD, and sea-of-gates. The diagrams in Figure 2.1 are meant to be suggestive of the general structure of each type of FPGA and no details are presented at this point. Instead, the major features possessed by FPGAs in each category are presented throughout this chapter, by describing commercially available chips from a total of ten companies. Some new architectural features for FPGAs have been suggested in recent research papers, but they are not described here [Kawa90] [Chow91] [Ebel91].

While the features offered in each company's product differ somewhat, a user's logic circuit can generally be implemented in any class of FPGA by making use of a set of sophisticated CAD tools. Some tools are developed specifically by the FPGA manufacturer, and others are offered through third-party vendors. The tools that are appropriate for each class of FPGA vary, but most of the steps that are required to implement a design are the same in each case. An example of the design flow used to implement a circuit in an FPGA is given at the end of this chapter.

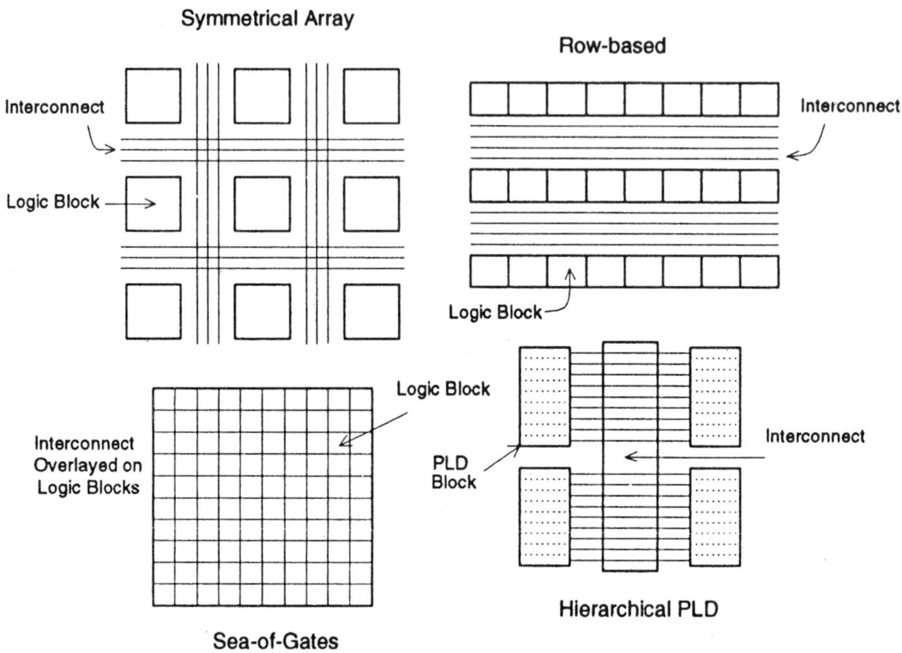

Figure 2.1 - *The Four Classes of Commercially Available FPGAs.*

2.1 Programming Technologies

Before discussing the features of commercially available FPGAs, it is useful to gain a better understanding of how these devices are made to be field-programmable. In Chapter 1, we used the word "switch" to refer to the entities that allow programmable connections between wire segments. We will continue to use this term throughout the book, but a more precise term for such an entity is *programming element*. Since there are a number of different ways of implementing a programming element, it has become customary to speak about the *programming technology* that is used to realize these elements. Programming technologies that are currently in use in commercial products are: static RAM cells, anti-fuses, EPROM transistors, and EEPROM transistors. While each of these technologies are quite different, the programming elements all share the property of being configurable in one of two states: ON or OFF.

The programming elements are used to implement the programmable connections among the FPGA's logic blocks, and a typical FPGA may contain more than 100,000 programming elements. For these reasons, the elements should have the following properties:

- the programing element should consume as little chip area as possible,
- the programming element should have a low ON resistance and a very high OFF resistance,
- the programming element should contribute low parasitic capacitance to the wiring resources to which it is attached, and
- it should be possible to reliably fabricate a large number of programming elements on a single chip.

Depending on the application in which the FPGA is to be used, it may also be desirable for the programming element to possess other features. For example, a programming element that is non-volatile might be attractive, as well as an element that is re-programmable. Re-programmable elements make it possible to re-configure the FPGA, perhaps without even removing it from the circuit board. Finally, in terms of ease of manufacture, it might be desirable if the programming elements can be produced using a standard CMOS process technology. The following sections describe each of the programming technologies in more detail. At the end, a table is presented that summarizes the characteristics of all of the programming technologies.

2.1.1 Static RAM Programming Technology

The static RAM programming technology is used in FPGAs produced by several companies: Algotronix, Concurrent Logic, Plessey Semiconductors, and Xilinx. In these FPGAs, programmable connections are made using pass-transistors, transmission gates, or multiplexers that are all controlled by SRAM cells. The use of a static RAM cell to control a CMOS pass-transistor is illustrated in Figure 2.2a. Alternatively, the RAM cell could control both the n-channel and p-channel transistors of a full transmission gate, or more than one RAM cell could be used to control the select inputs on a multiplexer. The later two options are illustrated by Figures 2.2b and 2.2c.

In the case of the pass-transistor approach in Figures 2.2a and 2.2b, the RAM cell controls whether the pass-gates are on or off. When off, the pass-gate presents a very high resistance between the two wires to which it is attached. When the pass gate is turned on, it forms a relatively low resistance connection between the two wires. For the multiplexer approach in Figure 2.2c, the RAM cells control which of the multiplexer's inputs should be connected to its output. This scheme would typically be used to optionally connect one of several wires to a single input of a logic block.

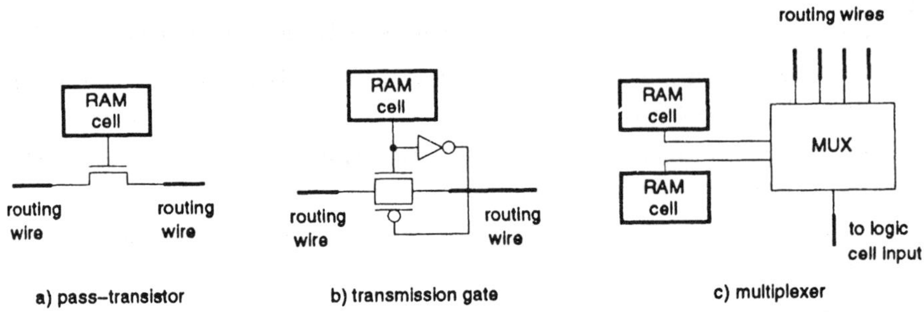

Figure 2.2 - *Static RAM Programming Technology.*

In an FPGA that uses the SRAM programming technology, the logic blocks may be interconnected using a combination of pass-gates and multiplexers. Since the static RAM is volatile, these FPGAs must be configured each time the power is applied to the chip. This implies that a system that includes such chips must have some sort of permanent storage mechanism for the RAM cell bits, such as a ROM or a disk. The RAM cells bits may be loaded into the FPGA either through a serial arrangement (if the RAM cells can be arranged in series during chip configuration) or each RAM cell may be addressed as an element of an array (as in a normal static RAM chip).

Compared with other programming technologies described in this section, the chip area required by the static RAM approach is relatively large. This is because at least five transistors are needed for each RAM cell, as well as the additional transistors for the pass-gates or multiplexers. The major advantage of this technology is that it provides an FPGA that can be reconfigured (in-circuit) very quickly and it can be produced using a standard CMOS process technology.

2.1.2 Anti-fuse Programming Technology

Anti-fuse programming technology is used in FPGAs offered by Actel Corp., QuickLogic, and Crosspoint Solutions. While the anti-fuse used in each of these FPGAs differs in construction, their function is the same. An anti-fuse normally resides in a high-impedance state but can be "fused" into a low-impedance state when programmed by a high voltage. In the remainder of this sub-section, the construction and use of the anti-fuses used by Actel and Quicklogic are described.

The Actel anti-fuse, called PLICE, is described in detail in [Hamd88]. It can be described as a square structure that consists of three layers: the bottom layer is composed of positively-doped silicon (n+ diffusion), the middle layer is a dielectric (Oxygen-Nitrogen-Oxygen insulator), and the top layer is made of poly silicon. This construction is illustrated by Figure 2.3a.

The PLICE anti-fuse is programmed by placing a relatively high voltage (18 V) across the anti-fuse terminals and driving a current of about 5 ma through the device. This procedure generates enough heat in the dielectric to cause it to melt and form a conductive link between the Poly-Si and n+ diffusion. Special high-voltage transistors are fabricated within the FPGA to accommodate the necessary large voltages and currents.

Both the bottom layer and top layer of the anti-fuse are connected to metal wires, so that, when programmed, the anti-fuse forms a low resistance connection (from 300 to 500 ohms) between the two metal wires. This arrangement is depicted in Figure 2.3b. The PLICE anti-fuse is manufactured by adding three specialized masks to a normal CMOS process.

The anti-fuse used by Quicklogic is called ViaLink. As described in [Birk91], it is similar to the PLICE anti-fuse in that it consists of three layers. However, a ViaLink anti-fuse uses one level of metal for its bottom layer, an alloy of amorphous silicon for its middle layer, and a second level of metal for the top layer. This structure is illustrated in Figure 2.4. When in the unprogrammed state, the anti-fuse presents over a gigaohm of resistance, but when programmed it forms a low-resistance path of about 80 ohms between the two metal wires. The anti-fuse is manufactured using three extra masks

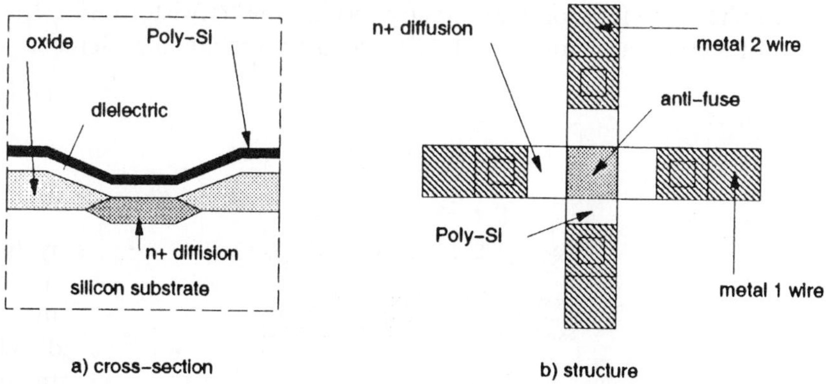

Figure 2.3 - *PLICE Anti-fuse Programming Technology.*

above a normal CMOS process. Here, a normal via is created for the anti-fuse, but the via is filled with the amorphous silicon alloy instead of metal.

The ViaLink anti-fuse is programmed by placing about 10 volts across its terminals. When sufficient current is supplied, this results in a change of state in the amorphous silicon and creates a conductive link between the bottom and top layers of metal.

The chip area required by an anti-fuse (either PLICE or ViaLink) is very small compared to the other programming technologies. However, this is somewhat offset by the large space required for the high-voltage transistors that are needed to handle the high programming voltages and currents. A disadvantage of anti-fuses is that their manufacture requires modifications to the basic CMOS process. Their properties are summarized in the table at the end of this section.

2.1.3 EPROM and EEPROM Programming Technology

EPROM programming technology is used in the FPGAs manufactured by Altera Corp. and Plus Logic. This technology is the same as that used in EPROM memories. Unlike a simple MOS transistor, an EPROM transistor comprises two gates, a *floating* gate and a *select* gate. The floating gate, positioned between the select gate and the transistor's channel, is so named because it is not electrically connected to any circuitry. In its normal (un-programmed) state, no charge exists on the floating gate and the transistor can be turned ON in the normal fashion using the select gate. However, when the transistor is programmed by causing a large current to flow between the source and drain, a charge is trapped under the floating gate. This charge has the effect of permanently turning the transistor OFF. In this way, the EPROM transistor can function as a programmable element. An

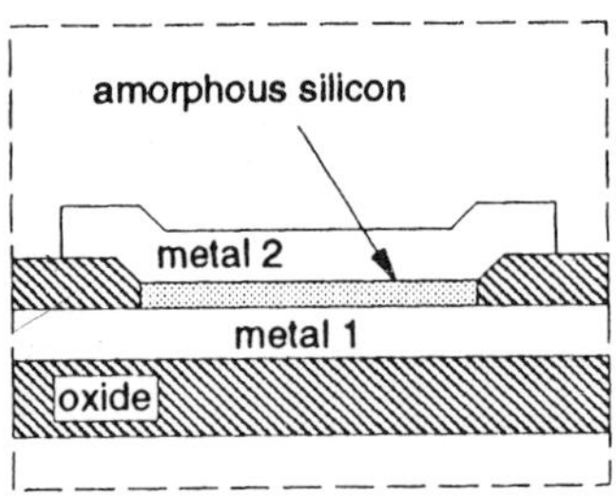

Figure 2.4 - *ViaLink Anti-fuse Programming Technology.*

EPROM transistor can be re-programmed by first removing the trapped charge from the floating gate. Exposing the gate to ultraviolet light excites the trapped electrons to the point where they can pass through the gate oxide into the substrate.

EPROM transistors are used in FPGAs in a different manner than are static RAM cells or anti-fuses. That is, rather than serving to programmably connect two wires, EPROM transistors are used as "pull down" devices for logic block inputs. This arrangement is illustrated in Figure 2.5. As the figure shows, one wire, called the "word line" (using memory terminology), is connected to the select gate of the EPROM transistor. As long as the transistor has not been programmed into the OFF state, the word line can cause the "bit line", which is connected to a logic block input to be pulled to logic zero. In both the Altera and Plus Logic FPGAs, many EPROM transistors, each driven by a different word line, are connected to the same bit line. Since a pull-up resistor is present on the bit line, this scheme allows the EPROM transistors to not only implement connections but also to realize wired-AND logic functions. A disadvantage of this approach is that the resistor consumes static power.

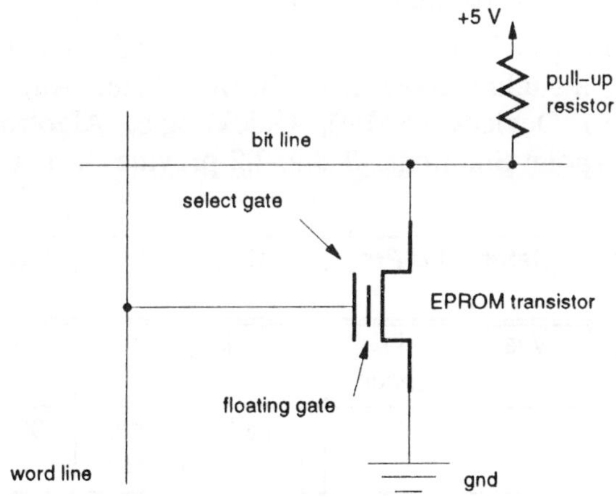

Figure 2.5 - *EPROM Programming Technology.*

One advantage of EPROM transistors is that they are re-programmable but do not require external storage. However, unlike static RAM, EPROM transistors cannot be re-programmed in-circuit.

The EEPROM approach (which is used in the FPGAs offered by Advanced Micro Devices) is similar to the EPROM technology except that EEPROM transistors can be re-programmed in-circuit. The disadvantage of using EEPROM transistors is that they consume about twice the chip area as

EPROM transistors and they require multiple voltage sources (for re-programming) which might not otherwise be required.

2.1.4 Summary of Programming Technologies

Table 2.1 lists some of the characteristics of the various programming technologies that have been discussed in this section. The second column from the left gives an indication of whether or not the programming element is volatile, and the third column states if the element is re-programmable. The amount of area consumed by the programming elements is indicated, in relative terms, in the fourth column. While each company that produces FPGAs may use a different base CMOS process, the numbers shown in the table are all normalized to 1.2 μm CMOS. The fifth and six columns list the series resistance of the programming element (when in the ON state) and the capacitance that the element adds to each wire to which it is attached, respectively. The resistance and capacitance numbers are approximate and are meant to provide a relative measure for the various elements.

2.2 Commercially Available FPGAs

This section provides descriptions of several commercially available FPGA families, including those from Xilinx, Actel, Altera, Plessey, Plus, Advanced Micro Devices (AMD), QuickLogic, Algotronix, Concurrent Logic, and Crosspoint Solutions. Table 2.2 provides a short summary, using

Programming Technology	Volatile	Re-Prog	Chip Area	R (ohm)	C (ff)
Static RAM Cells	yes	in circuit	large	1 - 2 K	10 - 20 ff
PLICE Anti-fuse	no	no	small anti-fuse, large prog. trans.	300 - 500	3 - 5 ff
ViaLink Anti-fuse	no	no	small anti-fuse, large prog. trans.	50 - 80	1.3 ff
EPROM	no	out of circuit	small	2 - 4 k	10 - 20 ff
EEPROM	no	in circuit	2x EPROM	2 - 4 k	10 - 20 ff

Table 2.1 - *Characteristics of Programming Technologies.*

terminology introduced at the beginning of this chapter, of several key features for each FPGA. It gives an indication of the FPGA's architecture, the type of logic block used, the structure of the routing resources, and the type of programming technology. Each of the devices in Table 2.2 is described in more detail in the following sections, with greater attention paid to the first three FPGAs in the table because they are the most widely used.

2.2.1 Xilinx FPGAs

The general architecture of Xilinx FPGAs is shown in Figure 2.6. It consists of a two-dimensional array of programmable blocks, called Configurable Logic Blocks (CLBs), with horizontal routing channels between rows of blocks and vertical routing channels between columns. Programmable resources are controlled by static RAM cells. There are three families of Xilinx FPGAs, called the XC2000, XC3000, and XC4000 corresponding to first, second, and third generation devices. Table 2.3 gives an indication of the logic capacities of each generation by showing the

Company	General Architecture	Logic Block Type	Programming Technology
Xilinx	Symmetrical Array	Look-up Table	Static RAM
Actel	Row-based	Multiplexer-Based	anti-fuse
Altera	Hierarchical-PLD	PLD Block	EPROM
Plessey	Sea-of-gates	NAND-gate	Static RAM
Plus	Hierarchical-PLD	PLD Block	EPROM
AMD	Hierarchical-PLD	PLD Block	EEPROM
QuickLogic	Symmetrical Array	Multiplexer-Based	anti-fuse
Algotronix	Sea-of-gates	Multiplexers & Basic Gates	Static RAM
Concurrent	Sea-of-gates	Multiplexers & Basic Gates	Static RAM
Crosspoint	Row-based	Transistor Pairs & Multiplexers	anti-fuse

Table 2.2 - *Summary of Commercially Available FPGAs.*

number of CLBs and an equivalent gate count, where a gate is ostensibly a 2-input NAND. The gate count measure is given in terms of "equivalent to a mask-programmable gate array of the same size." All FPGA manufacturers quote logic capacity by this measure, however the figures quoted by some are overly optimistic. The numbers given in Table 2.3, and in similar tables that appear later in this chapter are taken directly from the vendors, and should be interpreted accordingly. The design of the Xilinx CLB and routing architecture differs for each generation, so they will each be described in turn.

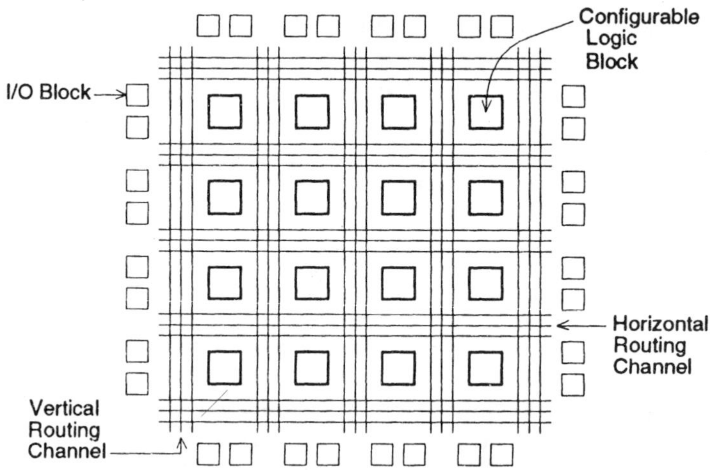

Figure 2.6 - *General Architecture of Xilinx FPGAs.*

Series	Number of CLBs	Equivalent Gates
XC2000	64 - 100	1200 - 1800
XC3000	64 - 320	2000 - 9000
XC4000	64 - 900	2000 - 20000

Table 2.3 - *Xilinx FPGA Logic Capacities.*

2.2.1.1 Xilinx XC2000

The XC2000 CLB, shown in Figure 2.7, consists of a four-input look-up table and a D flip-flop. The look-up table can generate any function of up to four variables or any two functions of three variables. Both of the CLB outputs can be combinational, or one output can be registered.

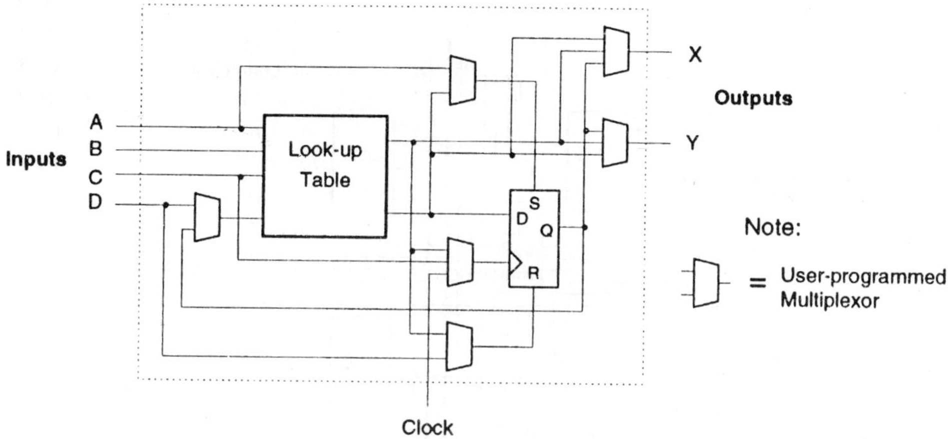

Figure 2.7 - *XC2000 CLB.*

As illustrated in Figure 2.8, the XC2000 routing architecture employs three types of routing resources: Direct interconnect, General Purpose interconnect, and Long Lines. Note that for clarity the routing switches that connect to the CLB pins are not shown in the figure. The Direct interconnect (shown only for the CLB marked with '*') provides connections from the output of a CLB to its right, top, and bottom neighbors. For connections that span more than one CLB, the General Purpose interconnect provides horizontal and vertical wiring segments, with four segments per row and five segments per column. Each wiring segment spans only the length or width of one CLB, but longer wires can be formed because each switch matrix holds a number of routing switches that can interconnect the wiring segments on its four sides. Note that a connection routed with the General Purpose interconnect will incur significant routing delays because it must pass through a routing switch at each switch matrix. Connections that are required to reach several CLBs with low skew can use the Long Lines, which traverse at most one routing switch to span the entire length or width of the FPGA.

2.2.1.2 Xilinx XC3000

The XC3000 is an enhanced version of the XC2000, featuring a more complex CLB and more routing resources. The CLB, as shown in Figure 2.9, includes a look-up table that can implement any function of five variables, or any two functions of four variables that use no more than five distinct inputs. The CLB has two outputs, both of which may be either combinational or registered.

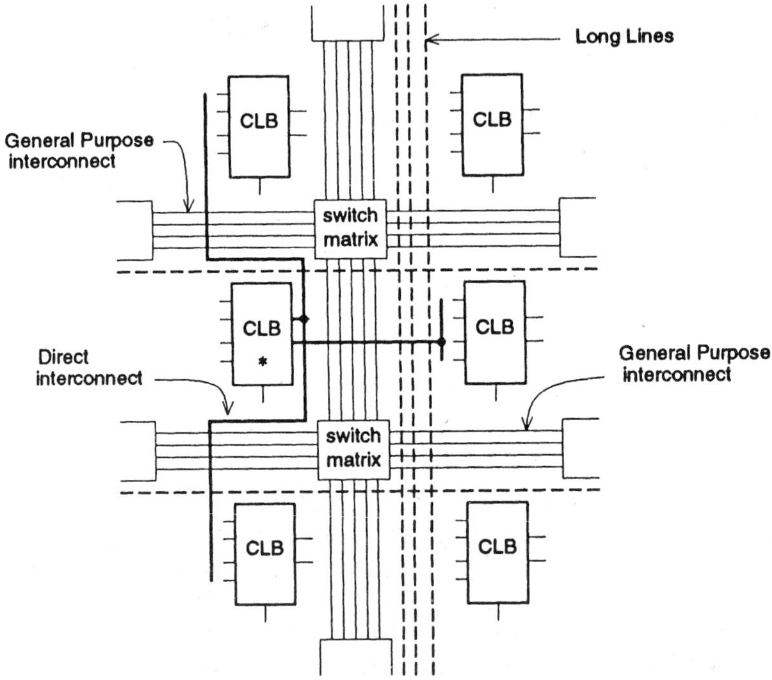

Figure 2.8 - *XC2000 Interconnect.*

Figure 2.10 shows that the XC3000 routing architecture is similar to that in the XC2000, having Direct interconnect, General Purpose interconnect, and Long Lines. Each resource is enhanced: the Direct interconnect can additionally reach a CLB's left neighbor, the General Purpose interconnect has an extra wiring segment per row, and there are more Long Lines.

The XC3000 also contains switch matrices that are similar to those in the XC2000. Figure 2.10 depicts the internal structure of an XC3000 switch matrix by showing, as an example, that the wiring segment marked with '*' can connect through routing switches to six other wiring segments. Although not shown in the figure, the other wiring segments are similarly connected, though not always to the same number of segments.

2.2.1.3 Xilinx XC4000

The XC4000 features several enhancements over its predecessors. The CLB, illustrated in Figure 2.11, utilizes a two-stage arrangement of look-up

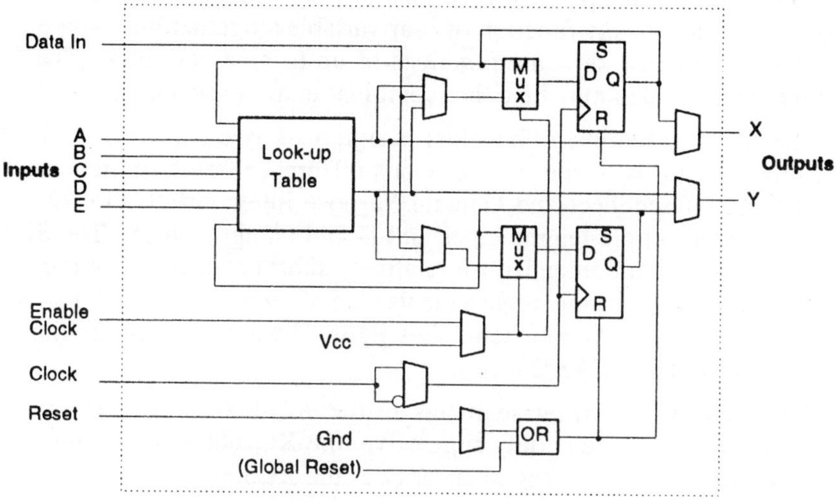

Figure 2.9 - *XC3000 CLB.*

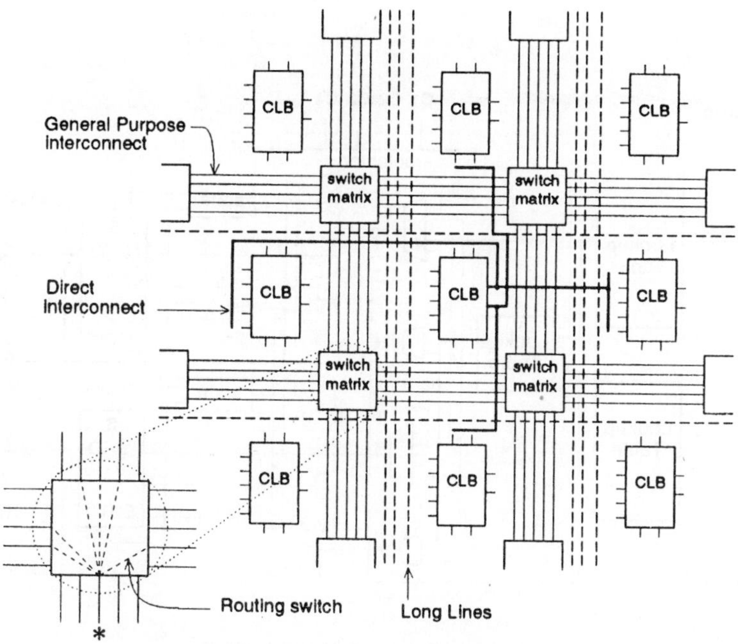

Figure 2.10 - *XC3000 Interconnect.*

tables that yields a greater logic capacity per CLB than in the XC3000. It can implement two independent functions of four variables, any single function of five variables, any function of four variables together with some functions of five variables, or some functions of up to nine variables. The CLB has two outputs, which may be either combinational or registered.

The XC4000 routing architecture is significantly different from the earlier Xilinx FPGAs, with the most obvious difference being the replacement of the Direct interconnect and General Purpose interconnect with two new resources, called Single-length Lines and Double-length Lines. The Single-length Lines, which are intended for relatively short connections or those that do not have critical timing requirements, are shown in Figure 2.12, where each X indicates a routing switch. This figure illustrates three architectural enhancements in the XC4000 series:

1. There are more wiring segments in the XC4000. While the number shown in the figure is only suggestive, the XC4000 contains more than twice as many wiring segments as does the XC3000.
2. Most CLB pins can connect to a high percentage of the wiring segments. This represents an increase in connectivity over the XC3000.

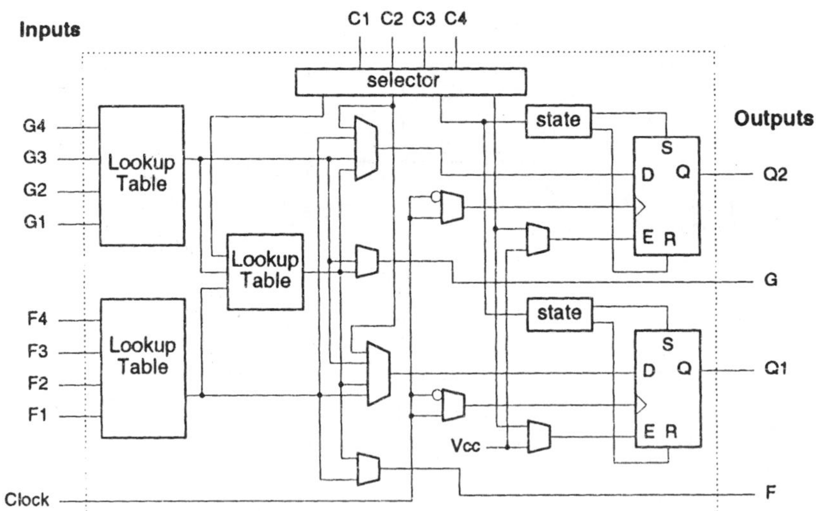

Figure 2.11 - *XC4000 CLB.*

3. Each wiring segment that enters a switch matrix can connect to only three others, which is half the number found in the XC3000.

It is interesting to note these three enhancements here because they are all supported by the architectural research that appears in Chapter 6 of this book.

The remaining routing resources in the XC4000, which includes the Double-length Lines and the Long Lines, are shown in Figure 2.13. As the figure shows, the Double-length Lines are similar to the Single-length Lines, except that each one passes through half as many switch matrices. This scheme offers lower routing delays for moderately long connections that are not appropriate for the low-skew Long Lines. For clarity, neither the Single-length Lines nor the routing switches that connect to the CLB pins are shown in Figure 2.13.

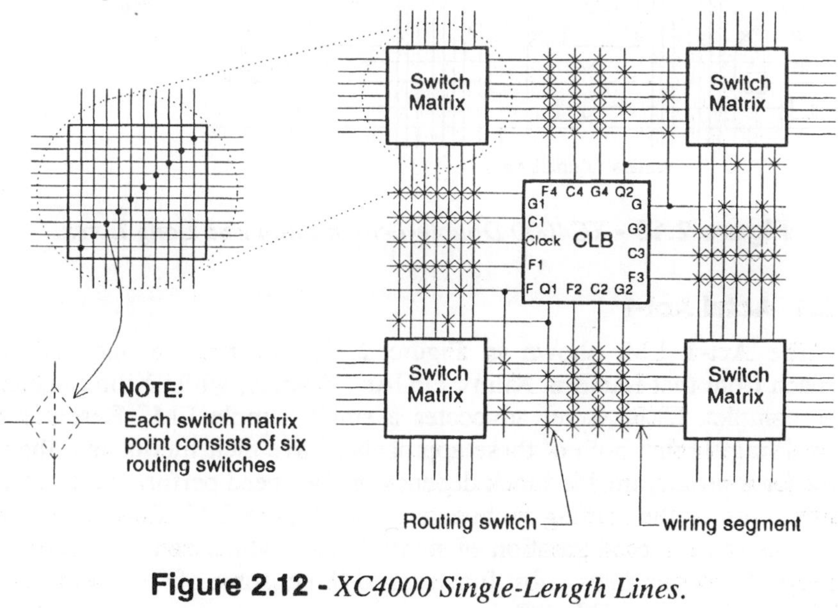

Figure 2.12 - *XC4000 Single-Length Lines.*

2.2.2 Actel FPGAs

The basic architecture of Actel FPGAs, depicted in Figure 2.14, is similar to that found in MPGAs, consisting of rows of programmable blocks, called Logic Modules (LMs), with horizontal routing channels between the rows. Each routing switch in these FPGAs is implemented by the PLICE anti-fuse that was described earlier in this chapter. Actel currently has two generations of FPGAs, called the Act-1 and Act-2, whose approximate logic capacities are shown in Table 2.4.

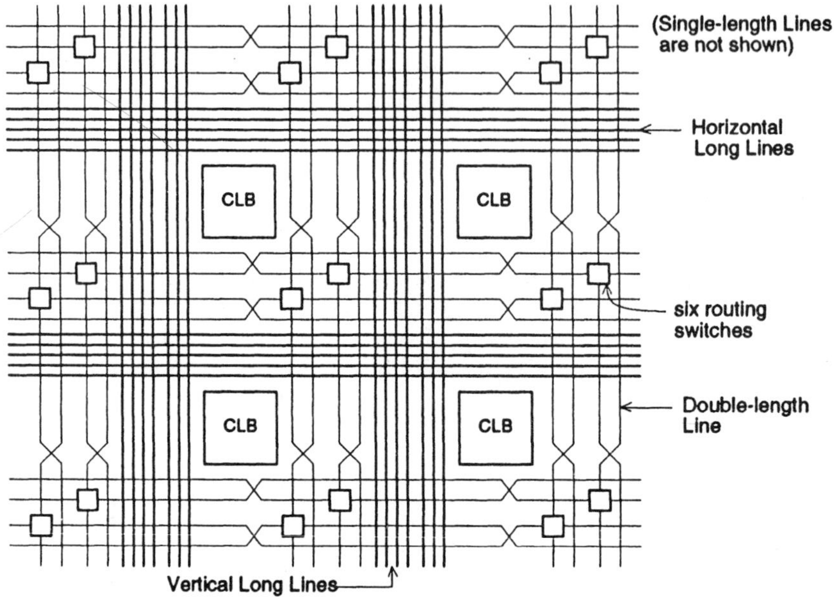

Figure 2.13 - *XC4000 Double-Length Lines and Long Lines.*

2.2.2.1 Actel Act-1

The Act-1 LM, shown in Figure 2.15, illustrates a very different approach from that found in Xilinx FPGAs. Namely, while Xilinx utilizes a large, complex CLB, Actel advocates a small, simple LM. Research has shown [Sing91] that both of these approaches have their merits, and the best choice for a programmable block depends on the speed performance and area requirements of the routing architecture. As Figure 2.15 shows, the Act-1 LM is based on a configuration of multiplexers, which can implement any function of two variables, most functions of three, some of four, up to a total of 702 logic functions [Mail90a].

The Act-1 routing architecture is illustrated in Figure 2.16, which for clarity shows only the routing resources connected to the LM in the middle of the picture. The Act-1 employs four distinct types of routing resources: Input segments, Output segments, Clock tracks, and Wiring segments. Input segments connect four of the LM inputs to the Wiring segments above the LM and four to those below, while an Output segment connects the LM output to several channels, both above and below the module. The Wiring segments consist of straight metal lines of various lengths that can be connected together through anti-fuses to form longer lines. The Act-1 features 22 tracks

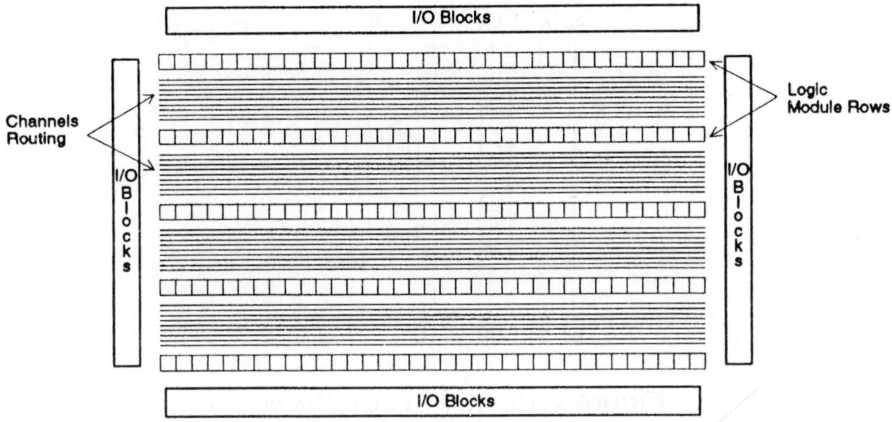

Figure 2.14 - *General Architecture of Actel FPGAs.*

Series	Number of LMs	Equivalent Gates
Act-1	295 - 546	1200 - 2000
Act-2	430 - 1232	6250 - 20000

Table 2.4 - *Actel FPGA Logic Capacities.*

of Wiring segments in each routing channel and 13 vertical tracks that lie directly on top of each LM column. Note that the figure shows only three vertical tracks lying on top of one LM column, but this is only to avoid cluttering of the diagram. Clock tracks are special low-delay lines that are used for signals that must reach many LMs with minimum skew.

2.2.2.2 Actel Act-2

The Act-2 device, an enhanced version of the Act-1, contains two different programmable blocks, called the C (Combinational) module and the S (Sequential) module. The C module is very similar to the Act-1 LM, although slightly more complex, while the S module is optimized to implement sequential elements (flip-flops).

The Act-2 routing architecture is also similar to that found in the Act-1. It features the same four types of routing resources, but the number of tracks is boosted to 36 in each routing channel and 15 in each column.

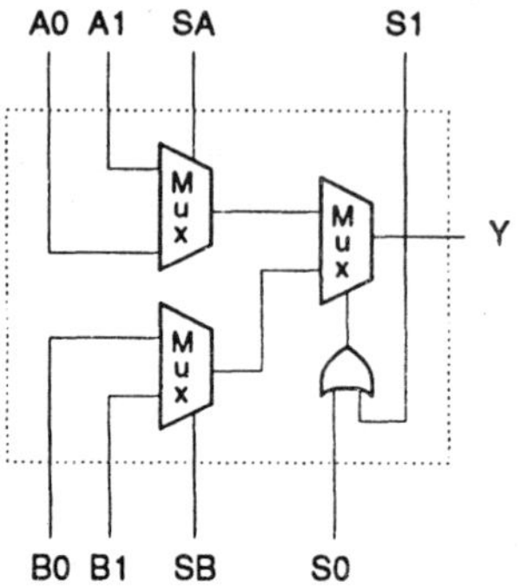

Figure 2.15 - *Act-1 Logic Module.*

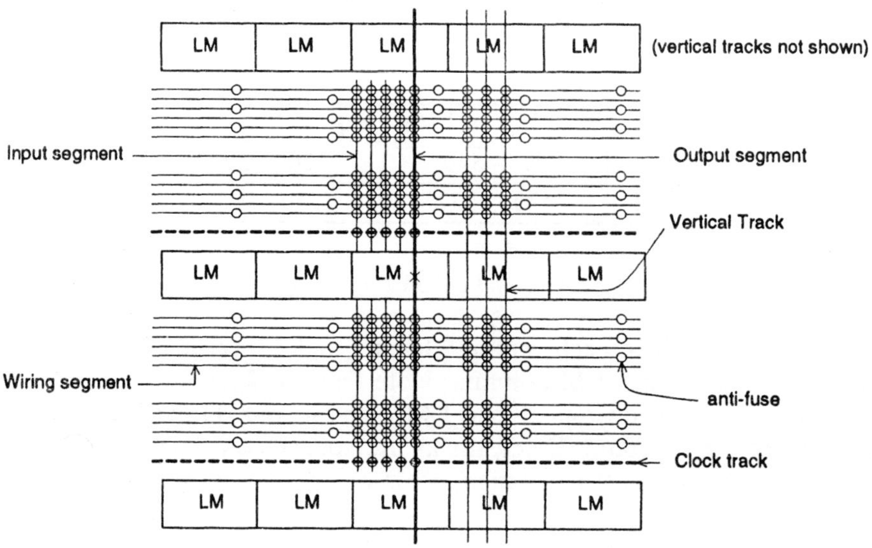

Figure 2.16 - *Act-1 Programmable Interconnect Architecture.*

2.2.3 Altera FPGAs

Altera FPGAs [Alt90] are considerably different from the others discussed so far because they represent a hierarchical grouping of Programmable Logic Devices. Nonetheless, they are FPGAs because they employ a two-dimensional array of programmable blocks and a programmable routing structure, they implement multi-level logic, and they are user-programmable.

Altera's general architecture, which is based on an EPROM programming technology, is illustrated in Figure 2.17. It consists of an array of large programmable blocks, called Logic Array Blocks (LABs), interconnected by a routing resource called the Programmable Interconnect Array (PIA). The logic capacities of the two generations of Altera FPGAs are listed in Table 2.5. We will focus on the EPM5000 series device here.

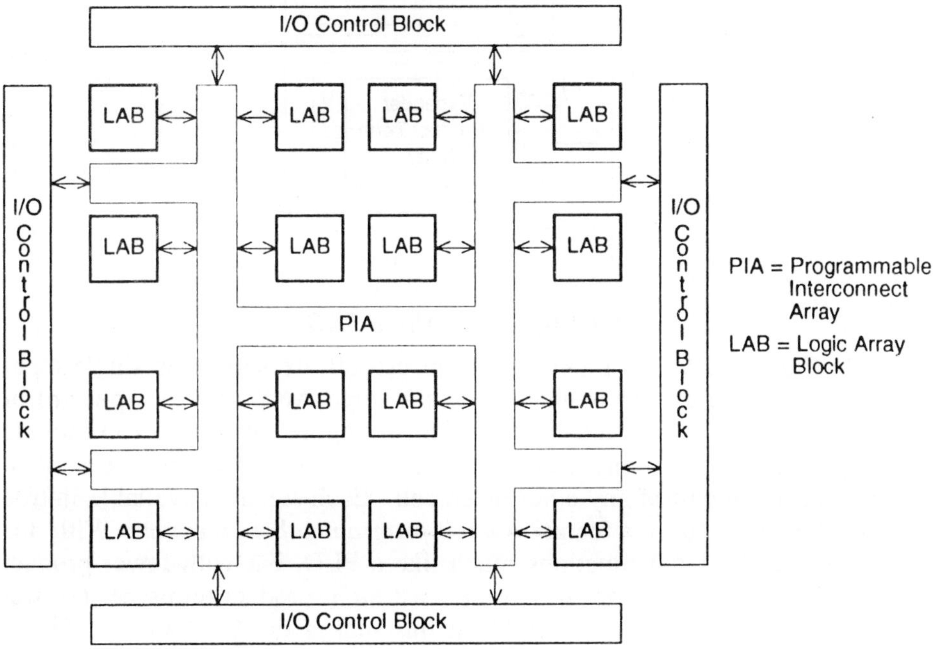

Figure 2.17 - *General Architecture of Altera FPGAs.*

Series	Number of LABs	Equivalent Gates
EPM5000	1 - 12	2000 - 7500
EPM7000	1 - 16	2000 - 20000

Table 2.5 - *Altera FPGA Logic Capacities.*

The Altera FPGA is a 2-level hierarchical grouping of logic blocks, called *macrocells*. The first level of the hierarchy, called a LAB, is illustrated in Figure 2.18. In addition to the macrocells, each LAB contains another kind of block, called the expander product terms. The number of macrocells in each LAB varies with the Altera device. As illustrated in Figure 2.19, each macrocell comprises three wide AND gates that feed an OR gate connected to an XOR gate, and a flip-flop. The XOR gate generates the macrocell output and can optionally be registered.

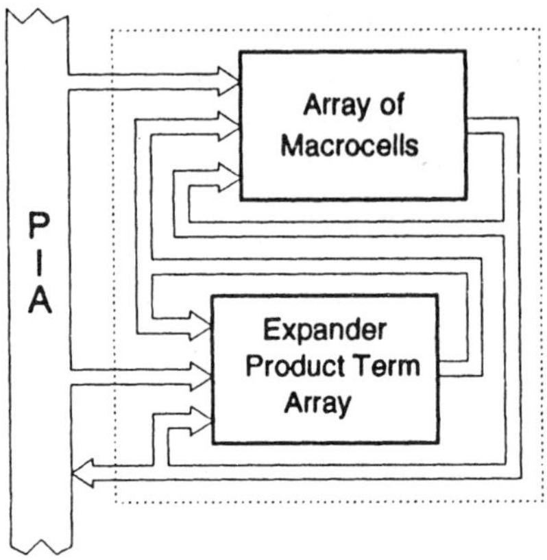

Figure 2.18 - *Altera LAB.*

In Figure 2.19, the inputs to the macrocell are shown as single-input AND gates because each is generated as a wired-AND (called a p-term) of the signals drawn on the left-hand side of the figure. A p-term can include any signal in the PIA, any of the LAB expander product terms (described below), or the output of any other macrocell. Each signal is available in true or complemented form, a feature we call *programmable inversion.* With this arrangement, the LAB functions much like a PLD, but with fewer product terms per register (there are usually at least eight product terms per register in a PLD). Altera claims [Alt90] that this makes the LAB more efficient because most logic functions do not require the large number of p-terms found in PLDs and the LAB supports wide functions by way of the expander product terms.

As illustrated in Figure 2.20, each expander product terms block consists of a number of p-terms (the number shown in the figure is only suggestive) that are inverted and fed back to the macrocells, and to itself. This arrangement permits the implementation of very wide logic functions because any macrocell has access to these extra p-terms.

The second level of the hierarchy provides connections among the LABs, which is accomplished through the PIA in Figure 2.17. The PIA consists of a number of long wiring segments that pass adjacent to every LAB. The PIA provides complete connectivity because each LAB input can be programmably connected to the output of any LAB, without constraints.

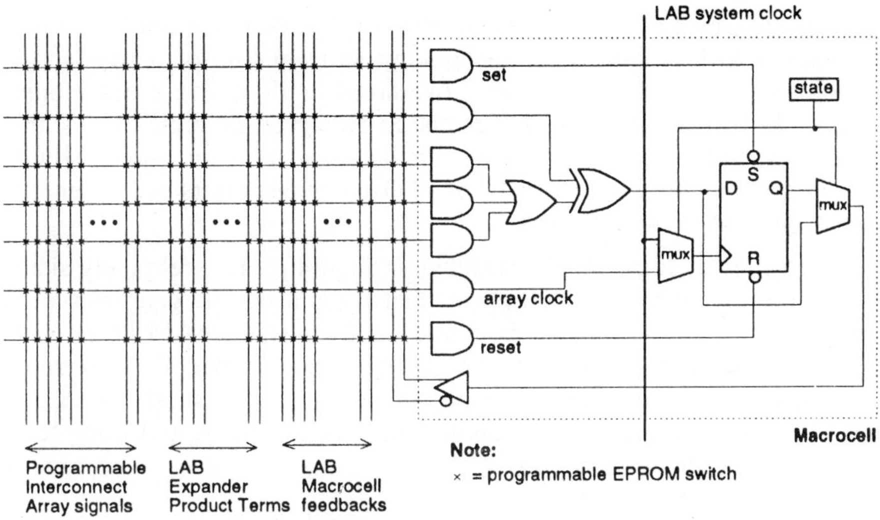

Figure 2.19 - *Altera Macrocell.*

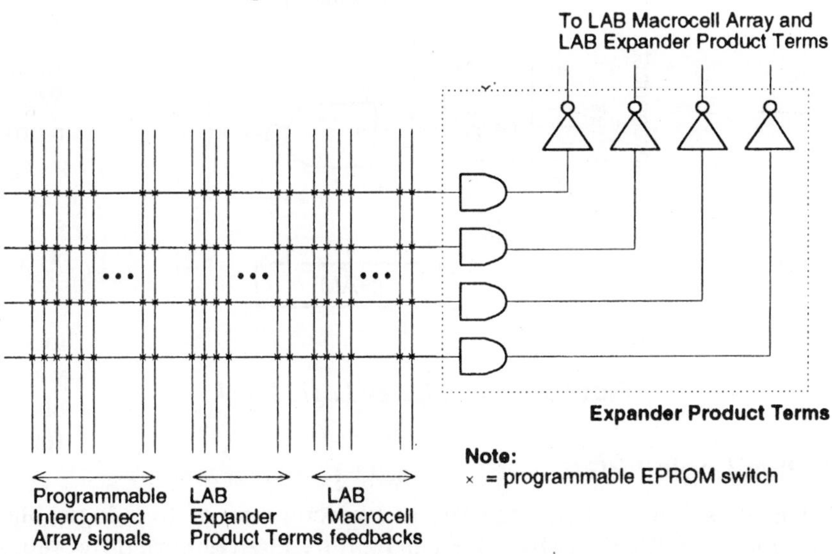

Figure 2.20 - *Altera Expander Product Terms.*

2.2.4 Plessey FPGA

The Plessey FPGA is called an Electrically Reconfigurable Array. It uses static RAM programming technology and consists of a regular two-dimensional array of logic blocks overlayed with a dense interconnect resource. With the routing resources placed on top of the logic blocks, these devices resemble the Sea-Of-Gates architecture used in some MPGAs. According to Plessey, their family of FPGA devices offers equivalent gate counts from 2000 to 40000 gates.

Each Plessey logic block, as shown in Figure 2.21, is relatively simple, containing an eight-to-two multiplexer that feeds a NAND gate, and a transparent latch. The multiplexer is controlled by a static RAM block and is used to connect the logic block to the routing resources, which comprise wiring segments of various lengths: Local interconnect for short connections, Short Range interconnect for moderate-length connections, and Long Range interconnect for long connections.

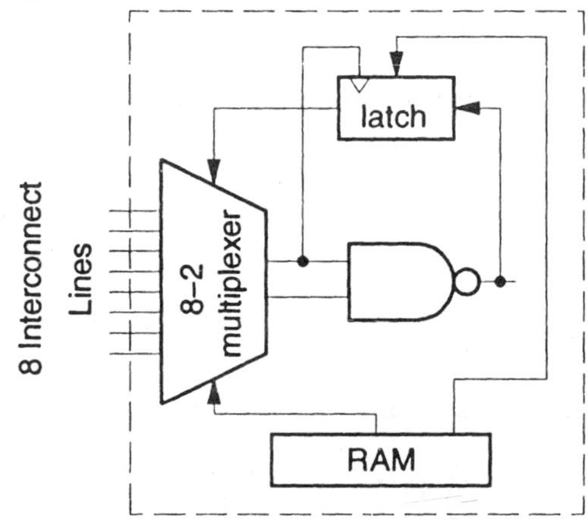

Figure 2.21 - *Plessey Logic Block.*

2.2.5 Plus Logic FPGA

The Plus Logic FPGA consists of two columns of four logic blocks, called Functional Blocks (FBs), that can be fully interconnected by a Universal interconnect Matrix (a full crossbar switch). The logic capacity of the Plus FPGA, which uses the EPROM programming technology, is 2000 to 4000 equivalent gates.

Compared to the first three FPGA architectures that were described in detail, this device is most like an Altera FPGA, but the FBs are even more

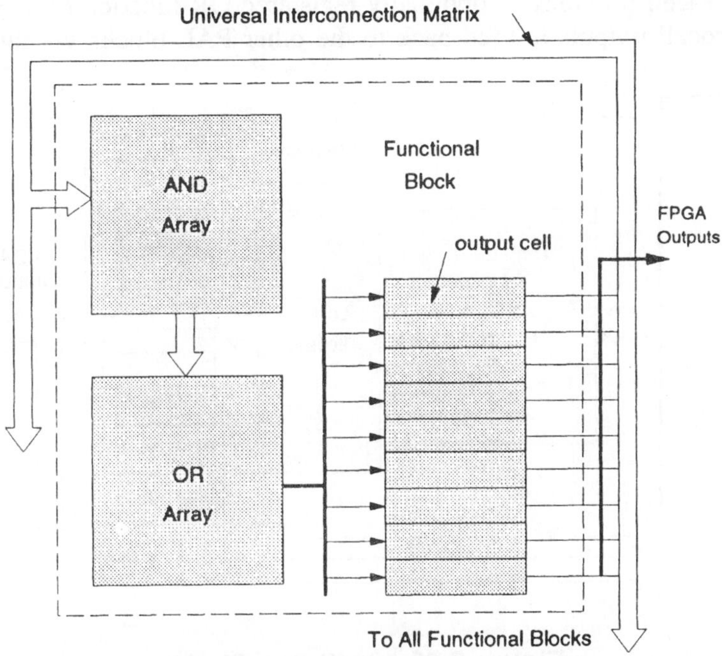

Figure 2.22 - *Plus Logic Functional Block.*

complex than the Altera LABs. Figure 2.22 depicts the structure of an FB. Each FB comprises a wide AND plane that feeds an OR plane, similar to a PLA device. The OR plane drives a third plane, which generates the nine (optionally registered) outputs of an FB. Each of these outputs is configurable to be any function of two terms from the OR array and one output of any other FB.

2.2.6 Advanced Micro Devices (AMD) FPGA

The AMD FPGA, based on EEPROM technology, can be considered to be an array of PAL devices that are interconnected by a switch matrix, similar to an Altera FPGA. The logic capacity of this FPGA varies from two to eight PAL blocks, or 900 to 3600 equivalent gates. Each PAL block consists of three main parts: an AND-plane, a Logic Allocator (LA), and a set of sixteen macrocells. The structure of the PAL block is shown in Figure 2.23. As the figure shows, the AND-plane feeds the LA, and the LA drives the macrocells. The LA distributes an appropriate number (the number is variable) of the product terms (p-terms) from the AND plane to individual macrocells. Each macrocell provides an optionally registered OR function of its p-terms. The macrocell outputs are fed back to the other PAL blocks via the switch matrix.

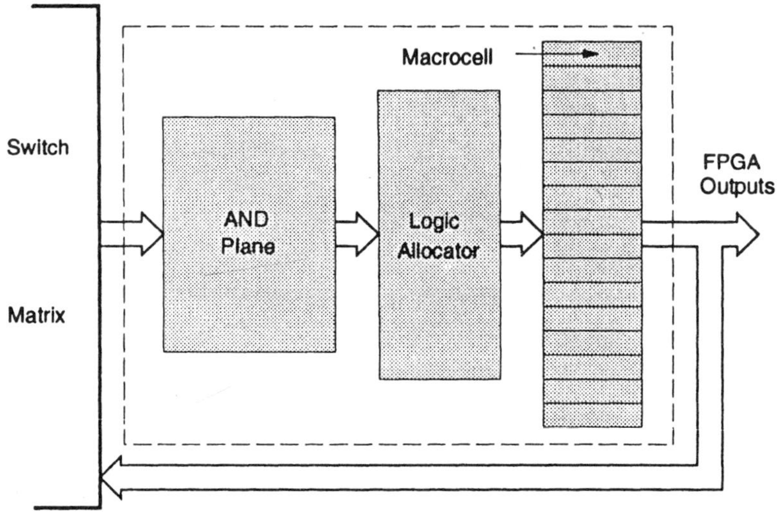

Figure 2.23 - *AMD PAL Block.*

2.2.7 QuickLogic FPGA

The QuickLogic FPGA [Birk91] consists of a regular two-dimensional array of blocks called pASIC Logic Blocks (pLBs). The logic capacities of the first generation of QuickLogic FPGAs is between 48 and 380 pLBs, or 500 to 4000 equivalent MPGA gates. QuickLogic claims that their logic capacity figures are more realistic than those quoted by other FPGA manufacturers. In terms of those other devices, according to [Birk91], the QuickLogic FPGAs have logic capacities of 1500 to 12000 gates.

The structure of each pLB is shown in Figure 2.24. It comprises four two-input AND gates feeding two two-input multiplexers, which feed a third multiplexer. The two first-stage multiplexers' select lines are driven by a six-input AND gate, and the second-stage multiplexer's select line is driven by another six-input AND gate. The second-stage multiplexer provides the block output, which can be optionally registered by a D flip-flop. In addition, the 6-input NAND gates can be used directly as pLB outputs.

The pLBs are interconnected by horizontal and vertical routing channels that provide full connectivity, in that every horizontal wire can be

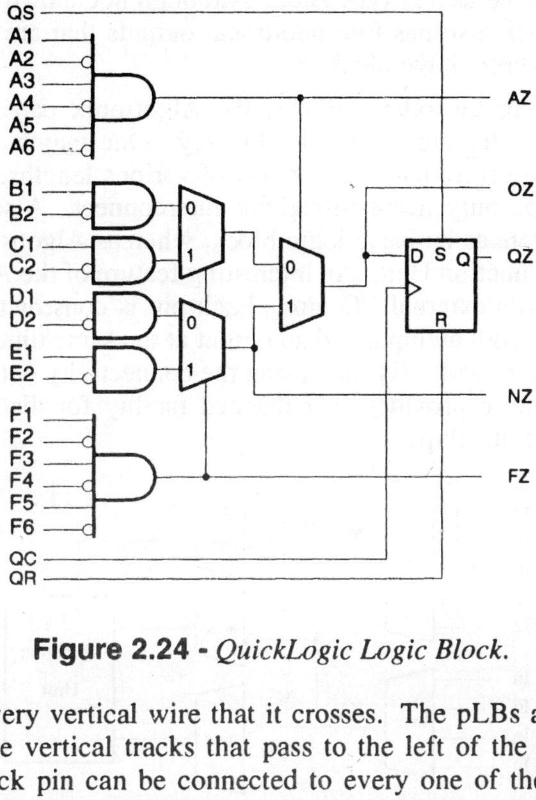

Figure 2.24 - *QuickLogic Logic Block.*

connected to every vertical wire that it crosses. The pLBs are only directly connected to the vertical tracks that pass to the left of the logic block, but every logic block pin can be connected to every one of these tracks. Programmed connections are formed in QuickLogic FPGAs using the ViaLink anti-fuse that was described in Section 2.1.2. As shown earlier, the ViaLink anti-fuse boasts very low ON resistance and parasitic capacitance. Compared with other FPGAs, the QuickLogic devices are most like those from Actel.

2.2.8 Algotronix FPGA

The Algotronix FPGA is organized as a 32×32 array of configurable blocks. Each block can only be directly connected to its four neighbors (left, right, above, below), but longer connections can be formed by routing signals through blocks (each block has special outputs that are available for this purpose). The Algotronix FPGA uses the static RAM programming technology, implementing connections with multiplexers. The devices have a logic capacity of 1024 programmable blocks, or 5000 equivalent gates.

As shown in Figure 2.25, the configurable blocks comprise several multiplexers that drive a Function Unit. The Function Unit is capable of implementing any logic function of its X1 and X2 inputs, or it can alternatively be configured as a D-type latch. Although not shown in the figure, the configurable block also has four additional outputs that are used for routing signals directly through the block.

Compared to the other FPGAs, the Algotronix part is most like the Sea-of-Gates architecture offered by Plessey. One major difference is that the Plessey architecture has interconnect of various lengths, while the Algotronix FPGA has only nearest-neighbor interconnect. Also, Plessey's part uses a NAND-gate as its basic logic block, whereas Algotronix employs the more complex Function Unit. An interesting feature of the Algotronix FPGA is the design of its external I/O pins. Each pin is constructed in a way that allows its use as both an input and an output at the same time (here, a 3-level, or ternary logic, is used). By increasing the connectivity between Algotronix FPGAs, this feature provides an enhanced facility for distributing a single design over multiple chips.

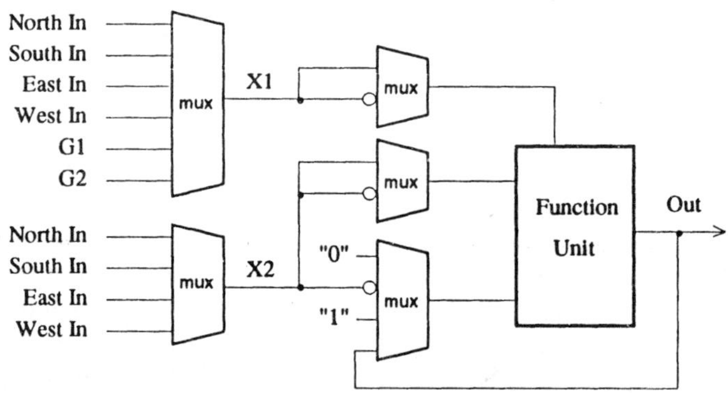

Figure 2.25 - *Algotronix Function Unit.*

2.2.9 Concurrent Logic FPGA

The CFA6006 FPGA, offered by Concurrent Logic, is based on a two-dimensional array of identical blocks, where each block is symmetrical on its four sides. The array holds 3136 of these blocks, providing a total logic capacity of about 5000 equivalent gates. Connections are formed using multiplexers that are configured by a static RAM programming technology.

The structure of the Concurrent Logic Block, shown in Figure 2.26, comprises user-configurable multiplexers, basic gates, and and a D-type flip-flop. The Concurrent FPGA is especially well suited for register-intensive and arithmetic applications since the Logic Block can easily implement a half-adder and a register bit. There are two direct connections (called A and

B) between each logic block and its four neighbors. Longer connections can be formed by routing signals through the multiplexers within the blocks (see Figure 2.26). Alternatively, although not shown in Figure 2.26, long connection can be implemented using a "bussing network", which can be viewed as wires of various lengths that are superimposed over the array of Logic Blocks.

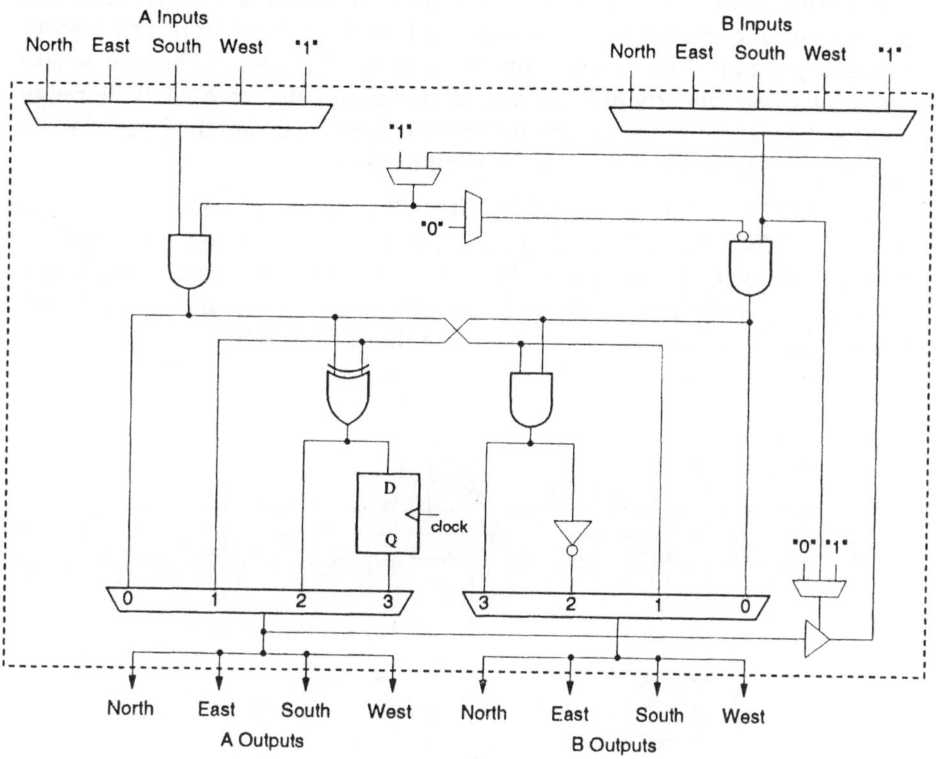

Figure 2.26 - *Concurrent Logic Block.*

2.2.10 Crosspoint Solutions FPGA

The Crosspoint FPGA [Marp92] differs from the others described thus far because it is configurable at the transistor level, as opposed to the logic block level. Basically, the architecture consists of rows of transistor pairs, where the rows are separated by horizontal wiring segments. Vertical wiring segments are also available, for connections among the rows. Each transistor row comprises two lines of series-connected transistors, with one line being NMOS and the other PMOS. The wiring resources allow individual transistor pairs to be interconnected to implement CMOS logic gates. The programming technology used for the programmable switches is similar to the ViaLink anti-fuse described earlier, in that it is based on amorphous silicon.

The structure of the transistor pair rows is illustrated by Figure 2.27. The figure shows the implementation of a NOR-gate and a NAND-gate using the transistor lines. As the figure indicates, transistor gates, sources and drains can be programmably interconnected to other transistors and also to power and ground. The series connection across the lines is broken where necessary by permanently holding a transistor in its OFF state. A wide range of logic gates can be implemented by the transistor lines and the interconnection patterns shown in the figure is only suggestive.

The FPGA currently offered by CrossPoint Solutions has a total logic capacity of 4200 gates. The chip has 256 rows of transistor pairs and an additional 64 rows of multiplexer-like structures not previously mentioned. With its row-based architecture, anti-fuse programming technology and multiplexers, the CrossPoint FPGA is most like those from Actel.

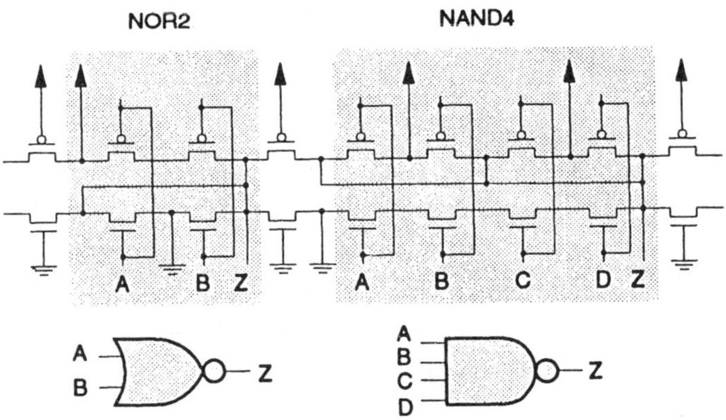

Figure 2.27 - *Crosspoint Transistor Pairs.*

2.3 FPGA Design Flow Example

This section indicates how a user's logic circuit can be implemented in an FPGA by describing the Xilinx design flow as an example. The Xilinx methodology has been selected because it provides a good example of the many steps involved in a typical FPGA CAD system. The design flow is depicted in Figure 2.28. As shown, the initial step in the process is the description of the logic circuit, which can be accomplished via a schematic capture tool or with Boolean expressions. This is followed by a translation that converts the original circuit description into a standard format used by the Xilinx CAD tools. The circuit is then passed through CAD programs that partition it into appropriate logic blocks (depending on which Xilinx part is being used), select a specific location in the FPGA for each logic block, and form the required interconnections. The performance of the implemented circuit can then be checked and its functionality verified. Finally, a bitmap is generated and can be downloaded in a serial fashion to configure the FPGA. Each of the steps from Figure 2.28 are described in more detail in the following sections.

2.3.1 Initial Design Entry

The description of the logic circuit can be entered using a schematic capture program. This involves using a graphical interface to interconnect circuit blocks. The available building blocks are taken from a component library. The library may be supplied by the vendor of the schematic capture program or by Xilinx itself, and is designed specifically for the Xilinx FPGA being used. As shown in Figure 2.28, schematic capture programs are offered by a wide variety of vendors.

An alternative way to specify the logic circuit is to use a Boolean expression or State Machine language. With this method, no graphical interface is involved. As Figure 2.28 shows, a number of different languages are available to support this option.

As the structure of Figure 2.28 indicates, it is also possible to use a mixture of the schematic capture and Boolean expression methods. The separate parts of the design are automatically merged after they are translated to the XNF format.

2.3.2 Translation to XNF Format

After the logic circuit has been fully designed and merged into one circuit, it is translated into a special format that is understood by Xilinx CAD tools. This format is called the Xilinx Netlist Format, or XNF. The translation utility is supported either by Xilinx or by the vendor of the logic entry tool. The translation process may also involve automatic optimizations of the circuit.

2.3.3 Partition

The XNF circuit is next *partitioned* into Xilinx Logic Cells. Note that the word partition is used by Xilinx, but the more common term for this step is *technology mapping*. Technology mapping converts the XNF circuit, which is a netlist of basic logic gates, into a netlist of Xilinx Logic Cells.

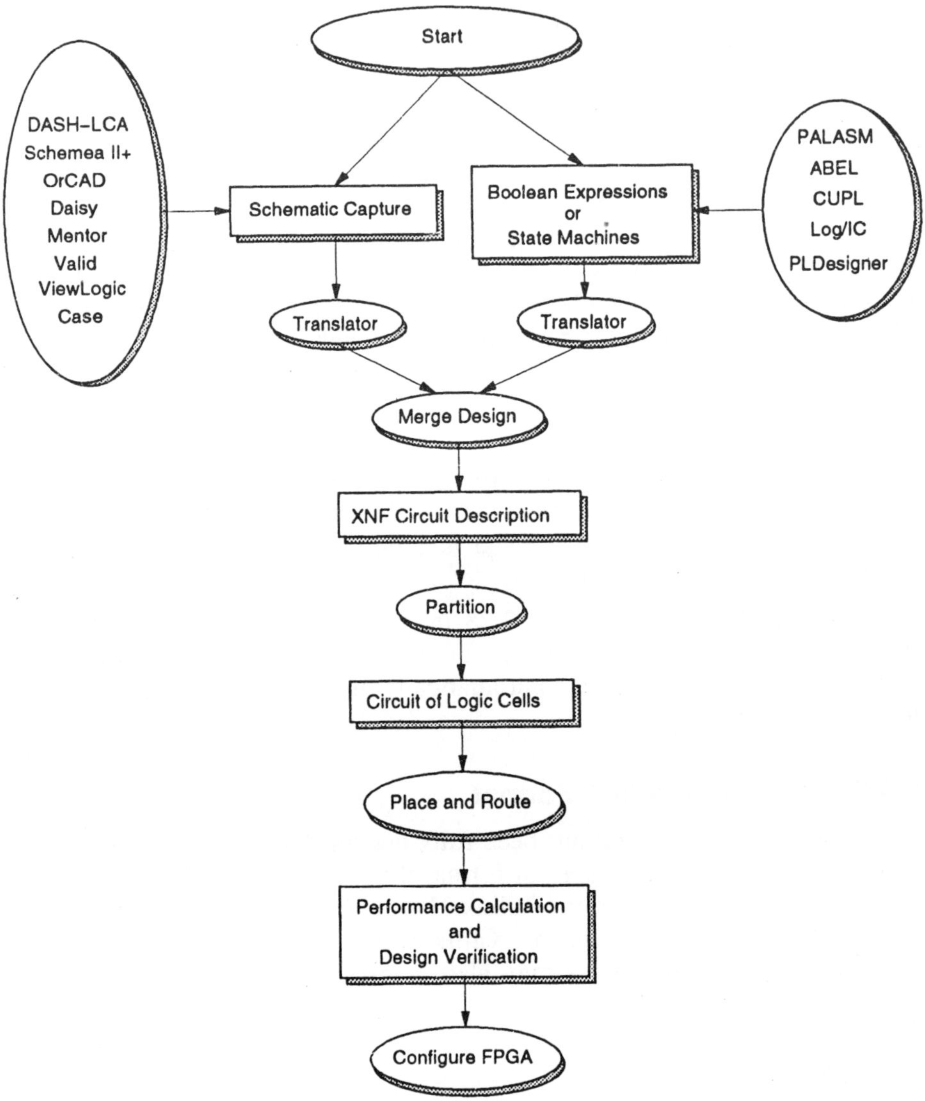

Figure 2.28 - *The Xilinx Design Flow.*

The Logic Cell used depends on which Xilinx product the circuit is to implemented in. The mapping procedure attempts to optimize the resulting circuit, either to minimize the total number of Logic Cells required, or the number of stages of Logic Cells in time-critical circuitry.

2.3.4 Place and Route

This step can be done automatically by CAD tools, manually by the user, or a mixture of the two. The first step is *placement*, in which each Logic Cell generated during the partitioning step is assigned to a specific location in the FPGA. Automatic placement is done using the Simulated Annealing algorithm [Sech87].

Following placement, the required interconnections among the Logic Cells must be realized by selecting wire segments and routing switches within the FPGA's interconnection resources. An automatic *routing* algorithm is provided for this task, based on a Maze Routing algorithm [Lee61].

Ideally, the placement and routing steps should be completely automatic, but in some cases manual assistance on the part of the user is required.

2.3.5 Performance Calculation and Design Verification

Once the circuit is routed, the physical paths of all signals within the FPGA are known. It is therefore possible to check the performance of the implementation, which can be done either by downloading the configuration bits into the FPGA and checking the part within its circuit board, or by using an interface to a timing simulation program. If the performance or functionality of the circuit is not acceptable, it will be necessary to modify the design at some point in the design flow. Once the timing and functionality is verified, the circuit implementation is complete.

2.4 Concluding Remarks

It should be clear by this point that a wide variety of FPGAs are commercially available. This chapter has described features of FPGAs from ten companies, without attempting to decide which architectural features are best. These issues are examined in the subsequent chapters.

CHAPTER 3 Technology Mapping for FPGAs

As mentioned in Chapter 1, the CAD system for a given FPGA performs several tasks needed to arrive at the final implementation of a circuit. This chapter focuses on the *logic synthesis* step in the CAD system, which consists of two separate phases called logic optimization and technology mapping. As illustrated in Figure 3.1, the original logic network is first manipulated by a *logic optimization* program, which produces an optimized network that is functionally equivalent to the original network. Logic optimization for FPGAs involves the same tasks as for other environments. Since a number of well-known logic optimization techniques have been described in several publications, we will only discuss it briefly in this chapter.

Following logic optimization, the *technology mapping* phase transforms the optimized network into a circuit that consists of a restricted set of circuit elements. In the FPGA environment, the circuit elements are the FPGA's logic blocks. This chapter examines the task of technology mapping in detail. It uses two types of FPGAs as examples: those that have logic blocks based on lookup tables (LUTs) and those that are based on multiplexers.

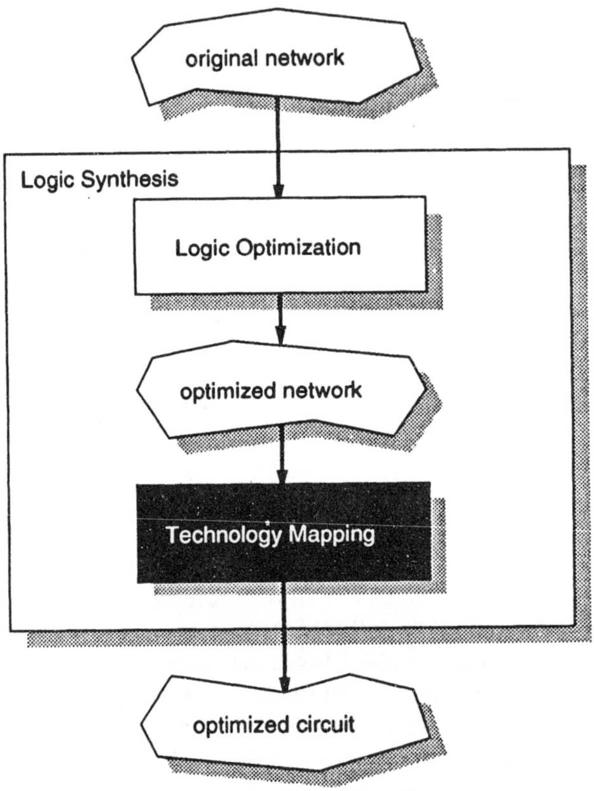

Figure 3.1 - *Logic Synthesis.*

3.1 Logic Synthesis

In this chapter, it is assumed that the original network to be manipulated by the logic synthesis tools is composed of a number of combinational functions. A network that contains flip-flops can also be processed, but it would first be broken up into a set of combinational functions at flip-flop boundaries. When read by the logic optimization phase, the network specifies a set of primary inputs that feed the combinational functions, and a set of primary outputs that are generated by the combinational functions.

One way to represent a set of combinational functions is to use a Directed Acyclic Graph (DAG) referred to as a Boolean network. An example of a small Boolean network is given in Figure 3.2. This network specifies a combinational function that has five primary inputs, *a*, *b*, *c*, *d*, *e*, and one

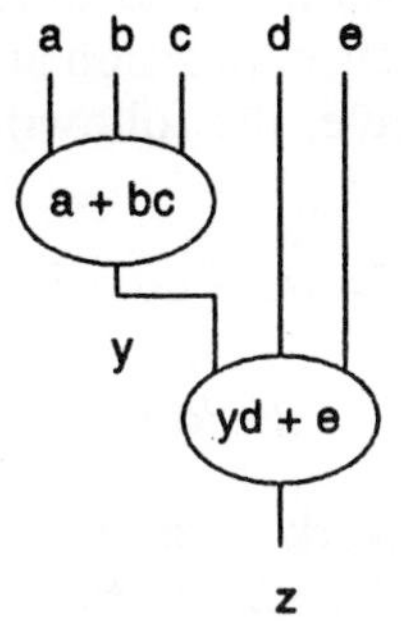

Figure 3.2 - *A Boolean Network.*

primary output, z. Each node in the Boolean network defines a function represented by a variable associated with the node. This function can be specified as a *local function* of the node's inputs, or as a *global function* of the network's primary inputs. For example, the node y in Figure 3.2 represents the local function $y = a + bc$, and the node z represents the global function $z = (a + bc)d + e$.

As Figure 3.1 illustrates, the end result of logic synthesis is an optimized circuit implementing the original Boolean network. The optimization goal is typically based on a measure of the area of the circuit, its speed performance, or both. The final circuit is also represented by a Boolean network. In the FPGA environment, each node represents an FPGA logic block and the local function of the node specifies the function implemented by the logic block.

3.1.1 Logic Optimization

The initial phase of many logic synthesis systems, for example misII [Bray87] and BOLD [Bost87], restructures the original network to reduce a cost function that is calculated directly from the network itself. The intention is to improve the final circuit by reducing the complexity of the network. Since this phase does not consider the type of element that will be used for the final circuit, it is called *technology-independent* logic optimization. The modifications applied to the network typically include redundancy removal and common subexpression elimination. Logic optimization may also exploit don't cares in the specification of the desired combinational function to simplify the network.

In the misII synthesis system the complexity of a network is measured by counting the number of *literals* in the local function for each node. Each local function is a sum-of-products expression and each instance of a variable counts as one literal. For example, the following 4-input, 2-output network has 11 literals

$$f = ac + ad + bc + bd$$

$$g = a + b + c$$

The complexity of this network can be reduced by the following modifications. The expression $(a + b)$ is factored out of the equations for nodes f and g, and a new node e, implementing the function $a + b$, is created. The variable e is then substituted back into the equations for nodes f and g, resulting is the following 7 literal network.

$$e = a + b$$

$$f = e(c + d)$$

$$g = e + c$$

3.1.2 Technology Mapping

After logic optimization has produced the optimized network, technology mapping transforms this network into the final circuit. This is done by selecting pieces of the network that can each be implemented by one of the available circuit elements, and specifying how these elements are to be interconnected. The circuit is optimized to reduce a cost function that typically incorporates measures of both area and delay. Conventional approaches to technology mapping have focused on using circuit elements from a limited set of simple gates, such as a Standard Cell library. The complex logic blocks used in FPGAs present difficulties for library-based approaches because they can each implement a large number of functions. The next section briefly discusses library-based technology mapping, and the following sections describe technology mapping programs developed specifically for lookup table and multiplexer-based FPGAs.

3.1.2.1 Library-Based Technology Mapping

An important advance in technology mapping for conventional technologies was the formalization introduced by Keutzer in DAGON [Keut87], and also used in misII [Detj87]. In this formalization, the set of available circuit elements is represented as a library of functions and the construction of the optimized circuit is divided into three subproblems: decomposition, matching and covering.

The original network is first decomposed into a canonical 2-input NAND representation. This decomposition guarantees that there will be no nodes in the network that are too large to be implemented by any library element, provided that the library includes a 2-input NAND. Note, however, that there are many possible 2-input NAND decompositions and that the one selected may not be the best decomposition.

After decomposition, the network is partitioned into a forest of trees. the optimal subcircuit covering each tree is constructed, and finally the circuit covering the entire network is assembled from these subcircuits. To form the forest of trees, the decomposed network is partitioned at fanout nodes into a set of single-output subnetworks. Each of these subnetworks is either a tree of a leaf-DAG. A leaf-DAG is a multi-input single-output DAG where only the input nodes have fanout greater than one. Each leaf-DAG is converted into a tree by creating a unique instance of every input node for each of its multiple fanout edges.

The optimal circuit implementing each tree is constructed using a dynamic programming traversal that proceeds from the leaf nodes to the root node. For every node in the tree an optimal circuit implementing the subtree extending from the node to the leaf nodes is constructed. This circuit consists of a library element that matches a subfunction rooted at the node and previously constructed circuits implementing its inputs. The cost of the circuit is calculated from the cost of the matched library element and the cost of the circuits implementing its inputs. To find the lowest-cost circuit, DAGON first finds all library elements that match subfunctions rooted at the node. The cost of the circuit using each of the candidate library elements is then calculated and the lowest cost circuit is retained. The set of candidate library elements is found by searching through the library and using tree matching [Aho85] to determine if each library element matches a subfunction rooted at the node.

As an example of the above procedure, consider the library shown in Figure 3.3 and the Boolean network in Figure 3.4. In this example, the library elements are standard cells and costs are given in terms of the area of the cells. The cost of the INV, NAND-2 and AOI-21 cells are 2, 3, and 4 respectively. In Figure 3.4a, the only library element matching at node E is the NAND-2 and the cost of the optimal circuit implementing node E is therefore 3. At node C the only matching library element is also the NAND-2. The cost of this NAND-2 is 3 and the cost of the optimal circuit

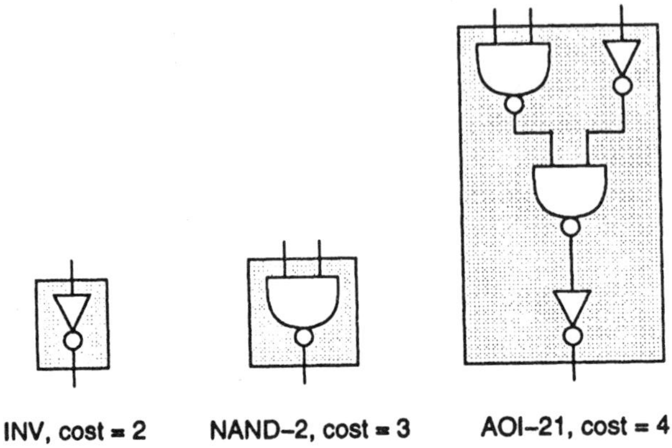

Figure 3.3 - *A Standard Cell Library.*

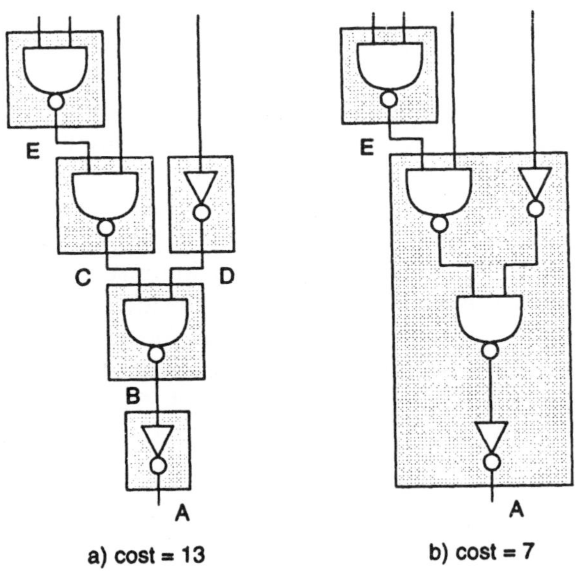

Figure 3.4 - *Mapping using Dynamic Programming.*

implementing its input E is also 3. Therefore, the cumulative cost of the optimal circuit implementing node C is 6.

Eventually, the algorithm will reach node A. For node A there are two matching library elements, the INV, as used in Figure 3.4a, and the AOI-21, as used in Figure 3.4b. The circuit constructed using the INV matching *A* includes a NAND-2 implementing node *B*, a NAND-2 implementing node *C*, an INV implementing node *D*, and a NAND-2 implementing node *E*. The cumulative cost of this circuit is 13. The circuit constructed using the AOI-21 matching *A* includes a NAND-2 implementing node *E*. The cumulative cost of this circuit is 7. The circuit using the AOI-21 is therefore the optimal circuit for implementing node *A*.

The tree matching algorithm requires that each library function be represented as a 2-input NAND decomposition. For some functions, however, there are many possible decompositions. The inclusion of all decompositions can significantly increase the size of the library and the computational cost of the matching algorithm.

3.2 Lookup Table Technology Mapping

Having introduced both logic optimization and technology mapping in the previous section, we will now consider technology mapping for lookup table circuits. Lookup tables (LUTs) are the basis of logic blocks in FPGAs from Xilinx, as described in Chapter 2. A K-input LUT is a digital memory that can implement any Boolean function of K variables. The K inputs are used to address a 2^K by 1-bit memory that stores the truth table of the Boolean function. The major obstacle in applying library-based technology mapping approaches to LUT circuits is the large number of different functions that a LUT can implement. A K-input LUT can implement 2^{2^K} different Boolean functions. However, the library representing a K-input LUT need not include all 2^{2^K} different functions, because input permutations, input inversions and output inversions can be used to reduce the number of functions in the library. For example there are 256 different 3-input functions, but considering input permutations there are only 80 different functions and considering input inversions and output inversions, there are only 14 different functions. However, the matching algorithms used in library-based technology mappers require the expansion of the library to include all possible decompositions of each function. For values of K greater than 3 the size of the library required to represent a K-input lookup table becomes impractically large. For this reason, new approaches to technology mapping are required for lookup table-based FPGAs.

There exist a number of LUT technology mappers, including: Chortle [Fran90] [Fran91a] [Fran91b], mis-pga [Murg90] [Murg91a] [Murg91b], Asyl [Abou90], Hydra [Filo91], Xmap [Karp91a] and VISMAP [Woo91a]. All of these programs map a Boolean network into a circuit of K-input LUTs, attempting to minimize either the total number of LUTs or the number of levels of LUTs in the final circuit. Minimizing the total number of LUTs allows the implementation of larger logic networks with the fixed number of lookup tables available in a given FPGA, and minimizing the number of levels improves the speed-performance of circuits. Chapter 4 discusses these issues at length. The following sections describe one algorithm, Chortle-crf, in detail and discuss the major features of the others.

3.2.1 The Chortle-crf Technology Mapper

Chortle-crf [Fran91a] maps a Boolean network into a circuit of K-input LUTs. Its objective is to minimize the number of LUTs in the final circuit. The nodes in the original network represent AND or OR functions, and inversion is identified by labelling edges. For example, in Figure 3.5a nodes a to m are the primary inputs of the network, and node z is the primary output. In this figure inverted edges are represented by a circle at the destination of the edge. The function specified for the primary output z is

$$z = (abc + def)(g + h + i)\overline{(jk + lm)}\,.$$

Figure 3.5b illustrates a circuit of 5-input LUTs implementing the Boolean network shown in Figure 3.5a. The dotted boundaries indicate the functions implemented by each LUT. The LUT y implements the Boolean function $y = jk + lm$ and the LUT z implements the Boolean function $z = x(g + h + i)\overline{y}$. Note that LUT y uses only 4 of the 5 available inputs. All examples in the remainder of this section will assume that the value of K is equal to 5.

The overall strategy used by Chortle-crf is similar to the library-based approach introduced by DAGON [Keut87]. The original network is first partitioned into a forest of trees and then each tree is separately mapped into a subcircuit of K-input LUTs. The final circuit is then assembled from the subcircuits implementing the trees.

The major innovation of Chortle-crf is that it simultaneously addresses the decomposition and matching problems using a bin-packing approximation algorithm. The correct decomposition of network nodes can reduce the number of LUTs required to implement the network. For example, consider the circuit of 5-input LUTs shown in Figure 3.6a. The shaded OR node is not decomposed, and four 5-input LUTs are required to implement the network. However, if the OR node is decomposed into the two nodes shown in Figure 3.6b, then only two LUTs are required. The challenge is to find the decomposition of every node in the network that minimizes the number of LUTs in the final circuit.

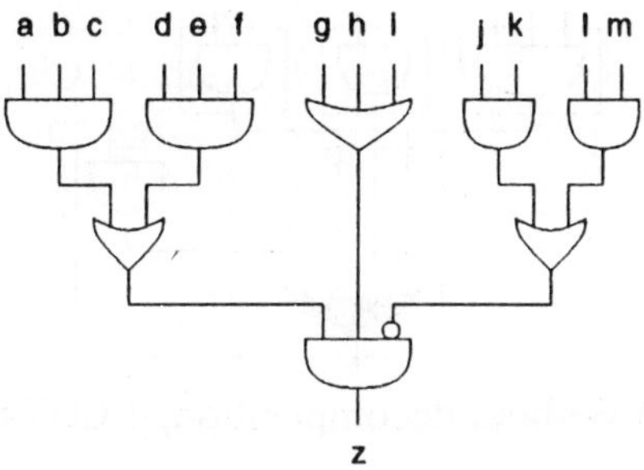

a) Boolean network

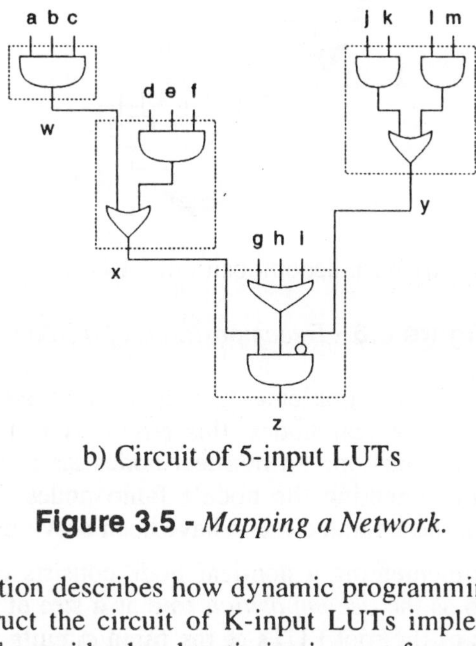

b) Circuit of 5-input LUTs

Figure 3.5 - *Mapping a Network.*

The next section describes how dynamic programming and bin packing are used to construct the circuit of K-input LUTs implementing each tree. Later sections will consider local optimizations at fanout nodes that further reduce the number of LUTs in the circuit by exploiting reconvergent paths and the replication of logic.

3.2.1.1 Mapping Each Tree

After the original network has been partitioned into a forest of trees, each tree is separately mapped into a circuit of K-input LUTs. Figure 3.7 outlines in pseudo-code the dynamic programming approach used to map each tree. The tree is traversed from its leaf nodes to its root node, and at

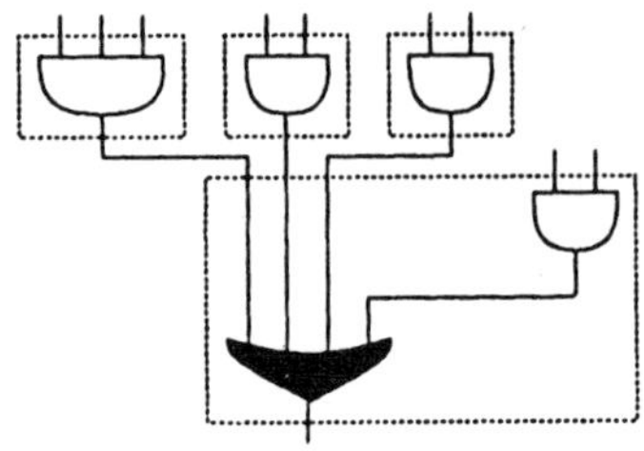

a) Without decomposition, 4 LUTs

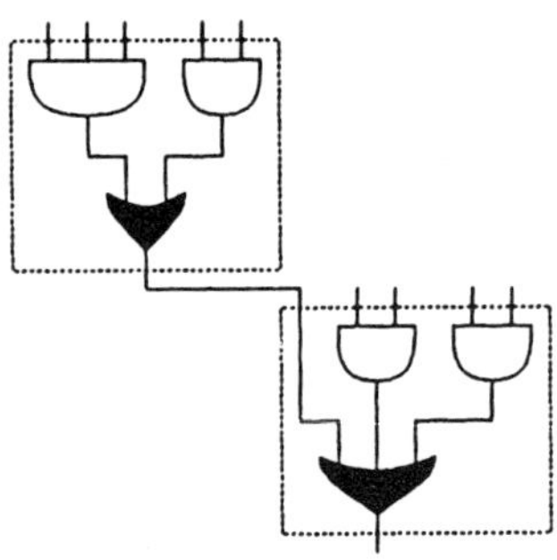

b) With decomposition, 2 LUTs

Figure 3.6 - *Decompositions of a Node.*

each node a circuit of LUTs implementing the subtree extending to the leaf nodes is constructed. For leaf nodes, this circuit is simply a single LUT implementing a buffer function. At non-leaf nodes the circuit is constructed from the circuits implementing the node's fanin nodes. The order of the traversal ensures that these fanin circuits have been previously constructed.

The circuit implementing a non-leaf node consists of two parts. The first part, referred to as the *decomposition tree*, is a tree of LUTs that implements the functions of the root LUTs of the fanin circuits and a decomposition of the non-leaf node. The second part is the non-root LUTs of the fanin circuits. For example, Figure 3.8a illustrates the circuits implementing the three fanin nodes of node *z*. The LUTs *w*, *x*, and *y* are the root LUTs of these fanin circuits and the LUTs *s*, *t*, *u*, and *v* are the non-root LUTs. Figure 3.8b illustrates the circuit implementing node *z* that is constructed from the fanin circuits. It includes the non-root LUTs *s*, *t*, *u*, and *v*, and the decomposition tree consisting of LUTs *w*, *z*. 1, and *z*. Note that the node *z* has been decomposed, and the new node *z*. 1 has been introduced.

```
MapTree (tree) {
    /* construct circuit implementing tree */

    traverse tree from leaves to root, at each node {
        if node is a leaf
            Circuit[node] ← single LUT buffering node
        else
            Circuit[node] ← MapNode (node)
    }
    return (Circuit[root])
}

MapNode (node) {
    /* construct circuit implementing node */

    /* separate fanin LUTs */
    faninLUTs ← root LUTs of Circuit [fanin] for all fanin nodes
    precedingLUTs ← non-root LUTs of Circuit [fanin] for all fanin nodes

    /* construct decomposition tree */
    decompositionTree ← DecomposeNode (node, faninLUTs)

    /* join decomposition tree and preceding LUTs */
    circuit ← decompositionTree ∪ precedingLUTs

    return (circuit)
}
```

Figure 3.7 - *Pseudo-code for Mapping a Tree.*

The essence of the dynamic programming approach is to construct the optimal circuit implementing each non-leaf node using the optimal circuits implementing its fanin nodes. The key to the algorithm is the definition of the optimal circuit. The principal optimization goal is to minimize the number of LUTs in the circuit, and the secondary goal is to minimize the number of inputs that the circuit's root LUT uses. This secondary goal is the key to ensuring that the optimal circuit implementing the non-leaf node is constructed from the optimal circuits implementing its fanin nodes.

Given that each fanin circuit contains the minimum number of LUTs, minimizing the number of LUTs in the decomposition tree minimizes the number of LUTs in the circuit implementing the non-leaf node. An important observation is that, for a given set of fanin circuits the number of LUTs in the best decomposition tree depends upon the number of inputs that the root LUT of each fanin circuit uses.

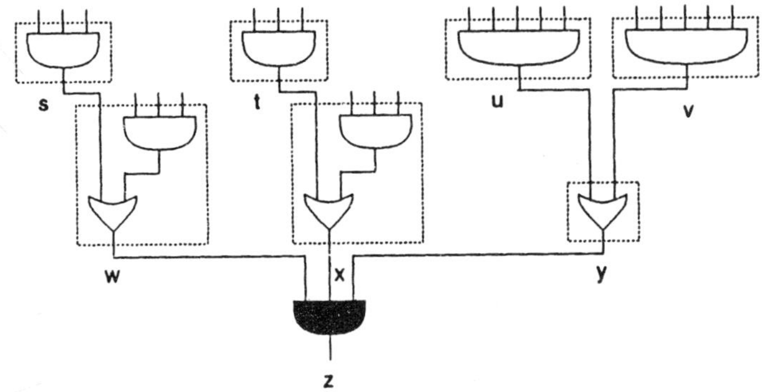

a) Fanin circuits

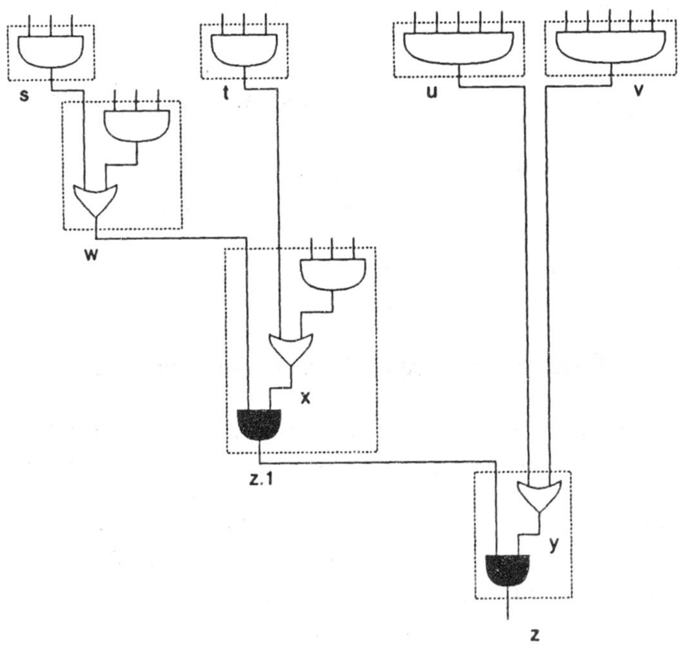

b) Circuit implementing node *z*

Figure 3.8 - *Mapping a Node.*

Consider two alternative circuits implementing one of the fanin nodes. Both alternatives contain the minimum number of LUTs, but the root LUT of the first one uses fewer inputs than the root LUT of the second one. The best decomposition tree constructed using the smaller root LUT may contain fewer LUTs than the best decomposition tree constructed using the larger root LUT. To ensure that the decomposition tree contains the minimum number of LUTs, the root LUT of each fanin circuit should use the minimum number of inputs. Therefore, the optimal circuit implementing each fanin node must contain the minimum number of LUTs and its root LUT must use the minimum number of inputs.

The dynamic programming approach requires that the optimal circuit implementing the non-leaf node satisfy the same optimization goals as the optimal circuits implementing the fanin nodes. Therefore the optimal circuit implementing the non-leaf node must also contain the minimum number of LUTs and its root LUT must use the minimum number of inputs. This requires that the decomposition tree contain the minimum number of LUTs and that its root LUT use the minimum number of inputs. The following section describes how the decomposition tree is constructed.

3.2.1.2 Constructing the Decomposition Tree

At each non-leaf node the decomposition tree implementing the functions of the root LUTs of the fanin circuits and a decomposition of the non-leaf node is constructed in two steps. The first step packs the root LUTs into what are called *second-level* LUTs. The second step connects these LUTs to form the complete decomposition tree.

Consider the node z and its fanin circuits shown in Figure 3.9a. This figure shows only the root LUTs of the fanin circuits, which are referred to as the fanin LUTs. Figure 3.9b shows the two-level decomposition constructed by the first step and Figure 3.9c shows the complete decomposition tree. Each second-level LUT implements some subset of the fanin LUTs and the corresponding decomposition of the node z. In Figure 3.9b the LUT $z.1$ implements the functions of the fanin LUTs u and v. In Figure 3.9c the output of LUT $z.1$ has been connected to an input of LUT $z.2$ and the output of LUT $z.2$ has been connected to an input of LUT z to form the complete decomposition tree. Note that in this example the fanin LUTs each implement a single AND gate, but in general the fanin LUTs can implement more complicated functions.

For a given set of fanin LUTs, the optimal decomposition tree contains the minimum number of LUTs and its root LUT uses the minimum number of inputs. The key to the construction of the optimal decomposition tree is

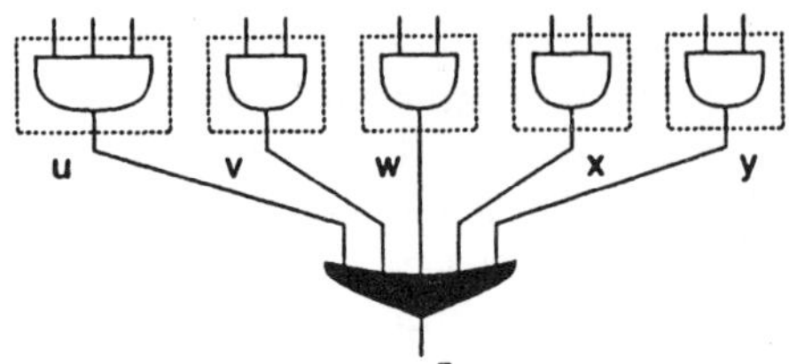

a) Fanin LUTs

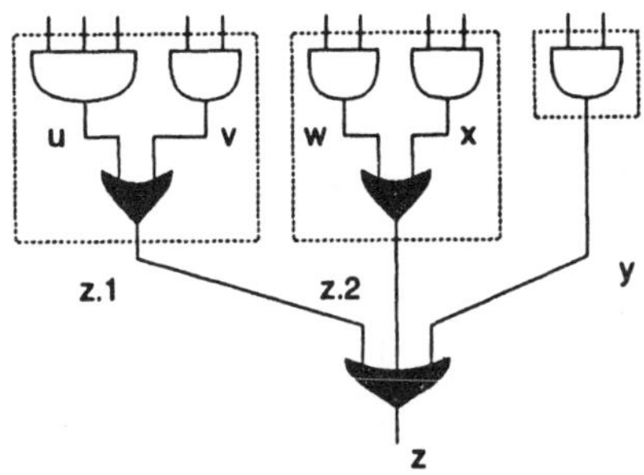

b) Two-level decomposition

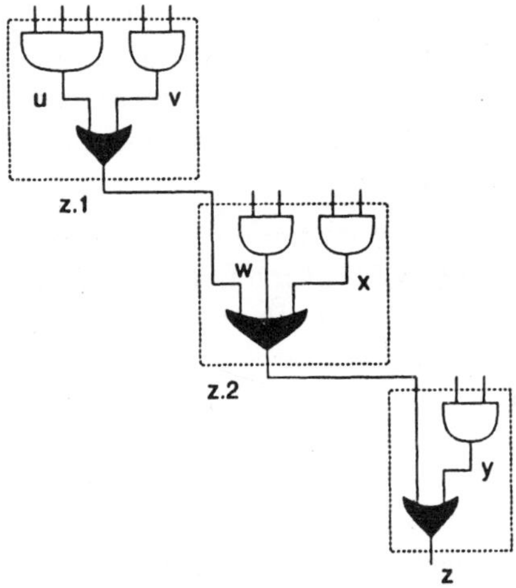

c) Multi-level decomposition

Figure 3.9 - *Constructing the Decomposition Tree.*

the construction of the two-level decomposition that contains the minimum number of LUTs. The major innovation of Chortle-crf is to approach the construction of two-level decomposition as a bin-packing problem. This approach is based on the observation that the function of each fanin LUT must be implemented completely within one LUT of the final decomposition tree.

In general, the goal of bin packing is to find the minimum number of subsets into which a set of items can be partitioned such that the sum of the sizes of the items in every subset is less than or equal to a constant C. The subsets can be viewed as a set of boxes packed into a bin of capacity C. In the construction of the two-level decomposition, the boxes are the fanin LUTs, and the bins are the second-level LUTs. The size of each box is its number of used inputs and the capacity of each bin is K. For example, in Figure 3.9a the boxes have sizes 3, 2, 2, 2, and 2. In Figure 3.9b the final packed bins have filled capacities of 5, 4, and 2.

Bin packing is known to be an NP-hard problem [Gare79], but there exist several effective approximation algorithms. The procedure used to construct the two-level decomposition, outlined as pseudo-code in Figure 3.10, is based on the First Fit Decreasing (FFD) algorithm. The fanin LUTs are referred to as boxes and the second-level LUTs are called bins. The procedure begins with an empty list of bins. The boxes are first sorted by size,

```
FirstFitDecreasing (node, faninLUTs) {
    /* construct two level decomposition */

    boxList ← faninLUTs sorted by decreasing size
    binList ← ∅

    while (boxList is not ∅) {
        box ← largest lookup table from boxList
        find first bin in binList such that size (bin) + size (box) ≤K

        if such a bin does not exist {
            bin ← a new lookup table
            add bin to end of binList
        }
        pack box into bin,
        /* implies decomposition of node */
    }
    return (binList)
}
```

Figure 3.10 - *Pseudo-code for Two-level Decomposition.*

and are then packed into bins one at a time, beginning with the largest box and proceeding in order to the smallest box. Each box is packed into the first bin in the list that has an unused capacity greater than or equal to the size of the box. If no such bin exists then a new bin is added to the end of the bin list and the box is packed into this new bin. Note that packing more than one box into a bin requires the introduction of a second-level decomposition node. For example, in Figure 3.9b when boxes *u* and *v* are packed into a bin this requires the introduction of the second-level decomposition node *z*. 1.

The procedure used to convert the two-level decomposition into the multi-level decomposition is outlined as pseudo-code in Figure 3.11. The second-level LUTs are first sorted by their size. Then, while there is more than one second-level LUT remaining, the output of the LUT with the greatest number of used inputs is connected to the first available unused input in the remaining LUTs. If no unused inputs remain then an extra LUT is added to the decomposition tree. Note that the decomposition node in the destination LUT is altered, and now implements part of the first level node.

```
DecomposeNode (node, faninLUTs) {
        /* construct tree of LUTs */
        /* implementing fanin LUTs and decomposition of node */

        /* construct two level decomposition */
        packedLUTs = FirstFitDecreasing (node, faninLUTs)
        lookList ← packedLUTs sorted by decreasing size

        while (lookList contains more than one lookup table) {
                src ← largest lookup table from lookList
                find first dst in lookList such that size (dst) + 1 ≤K

                if such a dst does not exist {
                        dst ← a new lookup table
                        add dst to end of lookList
                }
                connect src output to dst input,
                /* implies decomposition of node */
        }
        return (lookList)
}
```

Figure 3.11 - *Pseudo-code for Multi-level Decomposition.*

For example, in Figure 3.9c when LUT $z.1$ is connected to LUT $z.2$ the decomposition node $z.2$ is altered. This procedure constructs the optimal decomposition tree provided that the two-level decomposition contains the minimum number of LUTs and that its least filled LUT is as small as possible.

3.2.1.3 Optimality

The goal of Chortle-crf is to minimize the number of K-input LUTs required to implement the original Boolean network. The original network is first partitioned into a forest of trees and each of these is mapped separately. The final circuit implementing the original network is assembled from the subcircuits implementing the trees. For each tree, the subcircuit constructed by Chortle-crf is optimal provided that the value of K is less than or equal to 5 [Fran92]. For these values of K, the FFD bin-packing algorithm results in the two-level decomposition with the minimum number of LUTs and the smallest possible least filled LUT. This two-level decomposition leads to the optimal decomposition tree, which in turn leads to the optimal circuit implementing each non-leaf node, including the root node of the tree being mapped.

Even though the subcircuit implementing each tree in the forest is optimal, the final circuit implementing the entire network that is assembled from these subcircuits is not necessarily optimal. Partitioning the original network into a forest of trees precludes LUTs that realize functions containing reconvergent paths, and assembling the final circuit from the separate subcircuits implementing each tree precludes the replication of logic at fanout nodes. The following sections describe local optimizations that exploit reconvergent paths and the replication of logic at fanout nodes, to further reduce the number of LUTs in the final circuit.

3.2.1.4 Exploiting Reconvergent Paths

When the original Boolean network is partitioned at fanout nodes into single-output subnetworks, the resulting subnetworks are either trees or leaf-DAGs. In a leaf-DAG, a leaf node with out-degree greater than one is the source of reconvergent paths that terminate at some other node in the leaf-DAG. This section describes two alternative optimizations that exploit the reconvergent paths to improve the circuit implementing the terminal node. These optimizations replace the FFD algorithm and improve the two-level decomposition used to construct the decomposition tree. The first optimization uses an exhaustive search that repeatedly invokes the FFD algorithm. The second optimization uses a greedy heuristic that simplifies to the FFD algorithm when there are no reconvergent paths.

Both optimizations exploit reconvergent paths that begin at the inputs to the fanin LUTs and terminate at the non-leaf node being mapped. In the following description, the fanin LUTs are again referred to as boxes and the second-level LUTs are referred to as bins. Consider the set of boxes shown in Figure 3.12a. Two of the boxes share the same input, so there exists a pair of reconvergent paths that terminate at the shaded OR node. Each of these boxes has two inputs, for a total of four inputs. However, when they are packed into the same bin, as shown in Figure 3.12b, only 3 inputs are needed. The reconvergent paths are realized within the LUT and the total number of inputs used is less than the sum of the sizes of the two boxes. The decrease in the number of bin inputs that are used can allow additional boxes to be packed into the same bin and may therefore improve the final two-level decomposition. Figure 3.13a illustrates the two-level decomposition that can be constructed by applying the FFD bin-packing algorithm after the reconvergent paths have been realized within one LUT. By contrast, Figure 3.13b shows the result if the reconvergent paths are ignored, and the bin-packing algorithm is applied directly to the fanin LUTs. In this case, the two-level

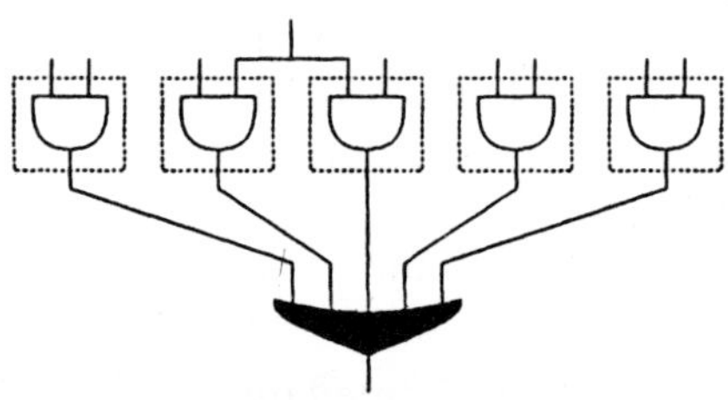

a) Fanin LUTs with shared inputs

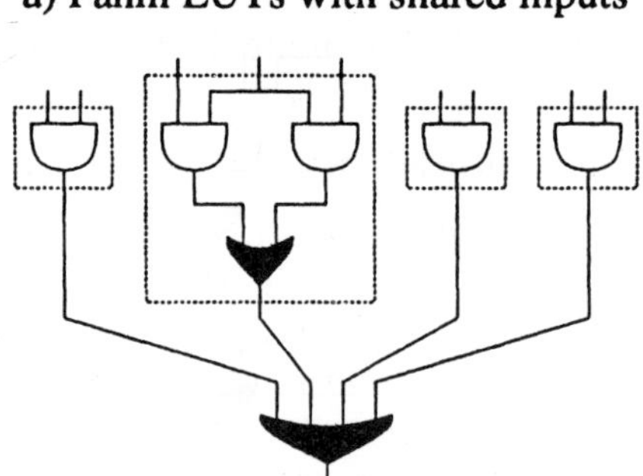

b) Reconvergent paths realized within one LUT

Figure 3.12 - *Local Reconvergent Paths.*

decomposition that realizes the reconvergent paths within a LUT contains fewer second-level LUTs.

The reconvergent paths can only be realized within one LUT if the two boxes with the shared input are packed into the same bin. To ensure that the boxes are packed together they can be merged before the FFD bin-packing algorithm constructs the two-level decomposition. However, forcing the two boxes into one bin can interfere with the FFD algorithm and actually produce an inferior two-level decomposition. To find the best two-level decomposition, the bin-packing algorithm is applied both with and without the forced merging of the two boxes and the superior two-level decomposition is retained.

When more than more pair of fanin LUTs share inputs, there are several pairs of reconvergent paths. To determine which pairs of reconvergent paths to realize within LUTs, an exhaustive search, outlined as pseudo-code in Figure 3.14, is used to find the best two-level decomposition. The

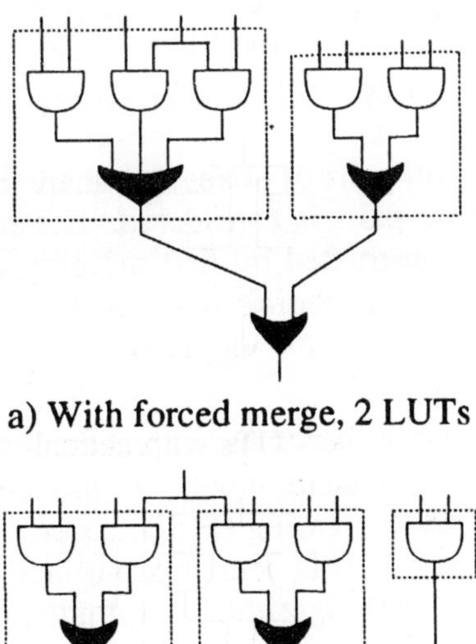

a) With forced merge, 2 LUTs

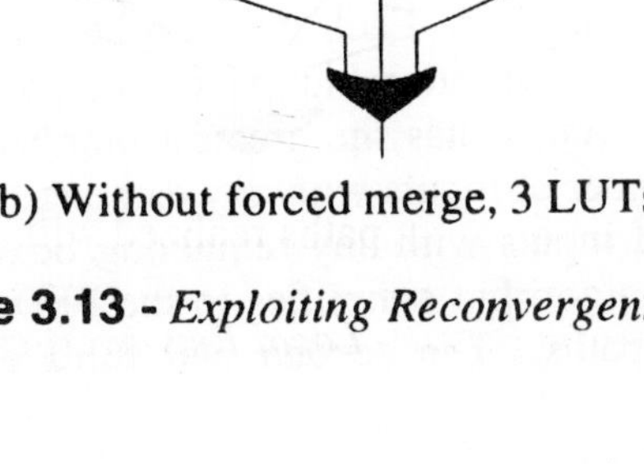

b) Without forced merge, 3 LUTs

Figure 3.13 - *Exploiting Reconvergent Paths.*

```
Reconverge (node, faninLUTs) {
    /* construct two level decomposition */
    /* exploit reconvergent paths, exhaustive search */

    pairList ← all pairs of faninLUTs with shared inputs
    bestLUTs ← ∅

    for all possible chosenPairs from pairList {
        mergedLUTs ← copy of faninLUTs with
          forced merge of chosenPairs
        packedLUTs ← FirstFitDecreasing (node, mergedLUTs)

        if packedLUTs are better than bestLUTs {
            bestLUTs ← packedLUTs
    }
    return (bestLUTs)
}
```

Figure 3.14 - *Pseudo-code for Reconvergent Optimization.*

search begins by finding all pairs of boxes that share inputs. Next, every possible combination of these pairs is considered. For each combination a two-level decomposition is constructed by first merging the respective boxes of the chosen pairs and then proceeding with the FFD bin-packing algorithm. The two-level decomposition with the fewest bins and the smallest least filled bin is retained.

The exhaustive search becomes impractical when there is a large number of pairs of boxes that share inputs. In this case, a heuristic, referred to as the Maximum Share Decreasing (MSD) algorithm, is used to construct the two-level decomposition. This heuristic, outlined as pseudo-code in Figure 3.15, is similar to the FFD algorithm, but it attempts to improve the two-level decomposition by maximizing the sharing of inputs when boxes are packed into bins. The MSD algorithm iteratively packs boxes into bins until all the boxes have been packed. Each iteration begins by choosing the next box to be packed and the bin into which it will be packed. The chosen box satisfies three criteria: first, it has the greatest number of inputs, second, it shares the greatest number of inputs with any existing bin, and third, it shares the greatest number of inputs with any remaining boxes. The first criterion ensures that the MSD algorithm simplifies to the FFD algorithm when there are no reconvergent paths. The second and third criteria encourage the

```
MaxShareDecreasing (node, faninLUTs) {
    /* construct two level decomposition */
    /* exploit reconvergent paths, greedy heuristic */

    boxList ← faninLUTs
    binList ← ∅

    while (boxList is not ∅) {
        box ← highest priority LUT from boxList
        /* criteria for highest priority box */
        /* 1) most inputs */
        /* 2) most inputs shared with a bin in binList */
        /* 3) most inputs shared with a box in boxList */

        find bin in binList that shares most inputs with boxLook

        if such a bin does not exist {
            bin ← a new LUT
            add bin to end of binList
        }
        pack box into bin exploiting shared inputs,
        /* implies decomposition of node */
    }
    return (binList)
}
```

Figure 3.15 - *Pseudo-code for Maximum Share Decreasing.*

sharing of inputs when the box is packed into a bin. The chosen box is packed into the bin with which it shares the most inputs while not exceeding the capacity of the bin. If no such bin exists then a new bin is created and the chosen box is packed into this new bin. Note that the second and third criteria only consider combinations of boxes and bins that will not exceed the bin capacity.

Both reconvergent optimizations only find local reconvergent paths that begin at the inputs of the fanin LUTs. However, when the fanin circuits are constructed no consideration is given to reconvergent paths that terminate at subsequent nodes. The propagation of these reconvergent paths through the fanin LUTs is dependent upon the network traversal order.

3.2.1.5 Replicating Logic at Fanout Nodes

This section describes how the replication of logic at fanout nodes can reduce the number of LUTs required to implement a Boolean network. Recall that the original Boolean network is partitioned into a forest of trees, and each tree is separately mapped into a circuit of K-input LUTs. When these separate circuits are assembled to form the circuit implementing the entire network, the replication of logic at fanout nodes can reduce the total number of LUTs in the final circuit. For example, in Figure 3.16a, three LUTs are required to implement the network when the fanout node is implemented explicitly as the output of a LUT. In Figure 3.16b, the AND gate implementing the fanout node is replicated and only two LUTs are required to implement the network.

When the original network is partitioned into a forest of trees, each fanout node is part of one source tree and several destination trees, as illustrated in Figure 3.17a. In this figure, the source and destination trees are represented by large triangles. The fanout node is the root of the source tree and it is a leaf in each of the destination trees.

The replication optimization considers replicating the function of the root LUT of the circuit implementing the source tree. In Figure 3.17a the

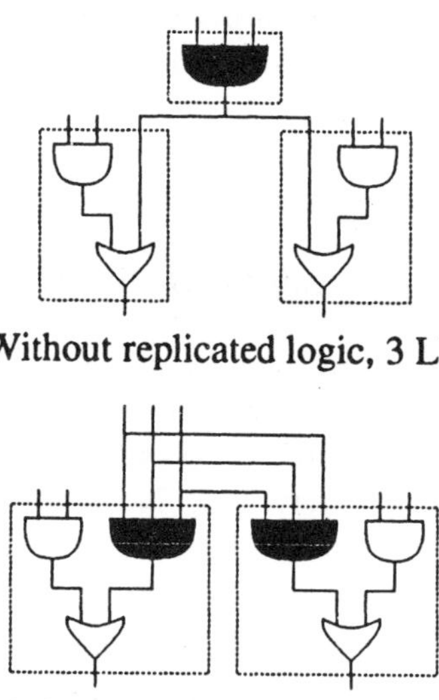

a) Without replicated logic, 3 LUTs

b) With replicated logic, 2 LUTs

Figure 3.16 - *Replication of Logic at a Fanout Node.*

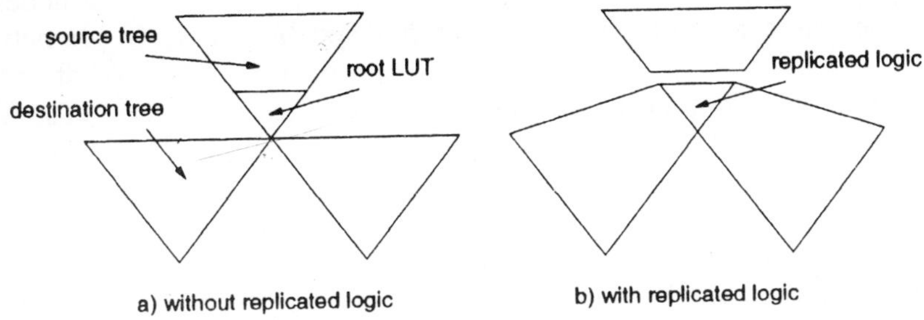

Figure 3.17 - *Replication of the Root LUT.*

small triangle at the root of the source tree represents the root LUT. The root LUT can be eliminated if a replica of its function is added to each of the destination trees, as illustrated in Figure 3.17b. If the total number of LUTs required to implement the destination trees does not increase, then eliminating the root LUT results is an overall reduction in the number of LUTs in the final circuit.

The replication optimization is outlined as pseudo-code in Figure 3.18. It begins by constructing the circuit implementing the source tree. The destination trees are first mapped without the replication of logic and are then remapped with a replica of the function of the source tree's root LUT added to each destination tree. If the total number of LUTs required to implement the destination trees with replication is less than or equal to the number without replication, then the replication is retained and the source tree's root LUT is eliminated.

When the original network contains many fanout nodes, the replication optimization is a greedy local optimization that is applied at every fanout node. If the destination tree of one fanout node is the source tree or destination tree of a different fanout node, there can be interactions between the replication of logic at the two fanout nodes. In this case, the replication of logic at the first fanout node may preclude replication at the second fanout node. The overall success of the replication optimization depends on the order in which it is applied to the fanout nodes.

3.2.1.6 Mapping into Xilinx 3000 CLBs

Chortle-crf can also map networks into circuits of Xilinx 3000 Configurable Logic Blocks (CLBs), which were described in Chapter 2. To map a network into a circuit of CLBs, Chortle-crf first maps it into a circuit

```
RootRep (srcTree) {
        /* decide if fanout LUT should be replicated */

        srcCircuit = mapTree (srcTree)
        rootLUT = root LUT of srcCircuit

        /* find cost without replication */
        noRepTotal = 0
        for all fanout dstTrees {
                noRepCircuit = mapTree (dstTree)
                noRepTotal = noRepTotal + number of LUTs in noRepCircuit
        }

        /* find cost with replication */
        repTotal = 0
        for all fanout dstTrees {
                add replica of rootLook to dstTree
                repCircuit = mapTree (dstTree)
                repTotal = repTotal + number of LUTs in repCircuit
        }

        if (repTotal ≤ noRepTotal) {
                retain repCircuits
                eliminate rootLUT from srcCircuit
        }
        else {
                retain noRepCircuits
        }
}
```

Figure 3.18 - *Pseudo-code for Root-LUT Replication.*

of 5-input LUTs and then assigns the functions specified by these LUTs to CLBs. Any single function can be assigned to a single CLB. In addition, any pair of functions that together use at most 5 distinct inputs, and that individually use at most 4 inputs, can be assigned to one CLB. To reduce the total number of CLBs in the final circuit, Chortle-crf maximizes the number of CLBs that implement a pair of functions using a Maximum Cardinality Matching approach, as introduced in mis-pga [Murg90].

3.2.1.7 Chortle-crf Performance

This section presents the results of using Chortle-crf to map 20 networks from the Microelectronic Center of North Carolina (MCNC) logic synthesis benchmark suite [MCNC91] into circuits of 5-input LUTs and Xilinx 3000 CLBs. Before the original networks are mapped, the number of literals in each network is reduced using the misII logic optimization program [Bray87]. The optimized networks are then mapped into circuits of 5-input LUTs using the following options:

- *-c*, without the reconvergent and replication optimizations
- *-cr*, with the reconvergent optimization
- *-cf*, with the replication optimization
- *-crf*, with both the reconvergent and replication optimizations

Column 1 in Table 3.1 lists the names of the networks, and columns 2 through 5 record the number of 5-input LUTs in the circuits that were constructed using the different options.

The table shows that the reconvergent and replication optimizations produce a 2.7 and 3.7 percent reduction, respectively, in the total number of 5-input LUTs. When combined, both optimizations produce a 14% reduction. This indicates an interaction between the reconvergent and replication optimizations. This occurs because the replication of logic at a fanout node can expose reconvergent paths, and thereby create additional opportunities for the reconvergent optimization.

Using the procedure outlined in the preceding section, the circuits of 5-input LUTs constructed using both the reconvergent and replication optimizations were converted into circuits of Xilinx 3000 CLBs. The results are shown in column 6. The final column in the table shows the execution time for Chortle-crf to construct the LUT circuits and convert them into the CLB circuits on a Sun 3/60.

3.2.2 The Chortle-d Technology Mapper

The objective of the Chortle-d technology mapper [Fran91b] is to improve circuit performance by minimizing the number of levels of LUTs in the final circuit. The overall approach used by Chortle-d is similar to that used in Chortle-crf, but a different procedure constructs the decomposition tree at each non-leaf node. Instead of minimizing the number of LUTs in the decomposition tree, Chortle-d minimizes its depth. Chortle-d is able to construct the optimal depth circuit of K-input LUTs implementing a network that is a tree, provided that the value of K is less than or equal to 6 [Fran92].

network	-c	-cr	-cf	-crf	-crf	Sun 3/60
	LUTs	LUTs	LUTs	LUTs	CLBs	sec.
5xp1	34	31	34	27	21	3.2
9sym	69	65	67	59	52	62.9
9symml	63	59	62	55	47	59.1
C499	166	164	158	74	50	15.9
C880	115	110	112	86	76	12.6
alu2	131	121	127	116	92	56.3
alu4	238	219	227	195	154	178.1
apex2	123	123	121	120	95	34.9
apex4	603	600	579	558	456	323.1
apex6	232	219	230	212	165	25.3
apex7	72	71	71	64	45	2.9
count	47	45	40	31	27	2.0
des	1073	1060	1050	952	789	291.9
duke2	138	136	126	120	90	9.1
e64	95	95	80	80	54	1.9
misex1	20	20	19	19	14	0.7
rd84	76	76	74	73	53	15.4
rot	219	207	208	189	133	14.0
vg2	24	24	23	21	19	0.6
z4ml	9	9	9	6	3	0.8

Table 3.1 - *Chortle-crf Results, 5-input LUTs and CLBs.*

The Chortle-d decomposition, when compared to the Chortle-crf decomposition, significantly increases the number of LUTs in the final circuit. To reduce this area penalty, Chortle-d uses the Chortle-crf decomposition on non-critical paths and uses a "peephole" optimization to eliminate single fanout LUTs that can be merged into their fanout LUTs.

Table 3.2 shows the results of using Chortle-d to map 20 MCNC networks into circuits of 5-input LUTs, and provides a comparison with Chortle-crf. Since the goal of Chortle-d is to reduce the number of levels of LUTs in the final circuit, the original networks are first optimized by misII to minimize the depth of the each network [Sing88]. The second and third columns in the table show the number of logic levels and the number of 5-input LUTs achieved by Chortle-crf, with both the reconvergent and replication optimizations. The fourth and fifth columns give the Chortle-d results.

network	Chortle-crf		Chortle-d	
	levels	LUTs	levels	LUTs
5xp1	4	27	3	27
9sym	8	65	5	57
9symml	7	62	4	54
C499	8	141	6	372
C880	13	172	8	309
alu2	13	128	9	219
alu4	17	231	10	495
apex2	8	124	6	184
apex4	11	624	5	1050
apex6	6	235	4	303
apex7	6	78	4	99
count	5	58	3	94
des	10	981	6	2243
duke2	7	152	4	225
e64	7	139	4	177
misex1	4	18	2	17
rd84	7	41	4	63
rot	11	214	6	302
vg2	5	39	4	51
z4ml	4	13	3	14

Table 3.2 - *Chortle-d Results, 5-input LUTs.*

The table shows that Chortle-d reduces the number of logic levels by 38 percent, but increases the number of 5-input LUTs by 79 percent.

3.2.3 Lookup Table Technology Mapping in mis-pga

The mis-pga technology mapper [Murg90] minimizes the number of K-input LUTs required to implement a Boolean network in two phases. The first phase decomposes the original network to ensure that every node can be implemented by a single K-input LUT, and the second phase uses a heuristic solution to a covering problem to reduce the number of LUTs in the final circuit.

In the first phase, four approaches are used to decompose nodes that cannot be implemented by a single LUT. The first approach is based on Roth-Karp decomposition [Roth62], the second approach is based on kernel extraction [Bray82], the third adapts the bin-packing approach introduced in Chortle-crf [Fran91a], and the fourth is based on Shannon cofactoring. On the basis of mapping 27 MCNC networks into circuits of 5-input LUTs [Murg91a], mis-pga requires 14% fewer LUTs than Chortle-crf, but it is 47 times slower.

Mis-pga addresses technology mapping for Xilinx 3000 CLBs by first mapping the Boolean network into 5-input functions and then assigning these functions to CLBs. Each CLB can implement one 5-input function or two functions of up to 4 inputs that together have no more than 5 inputs. Maximizing the number of CLBs that implement two functions, and thereby minimizing the total number of CLBs, is restated and solved as a Maximum Cardinality Matching problem.

Mis-pga also includes optimizations that improve speed-performance by reducing the number of levels of LUTs in the final circuit [Murg91b]. The original network is first decomposed into a depth-reduced network of 2-input nodes [Sing88] and then the critical paths are traversed from the primary inputs to the primary outputs. A critical node at depth d is collapsed into its fanout nodes, at depth $d+1$, whenever the resulting node is feasible, or can be redecomposed with a reduction in depth. Compared to Chortle-d, mis-pga requires 6% more levels of 5-input LUTs, but uses 31% fewer LUTs to implement 27 MCNC networks [Murg91b].

3.2.4 Lookup Table Technology Mapping in Asyl

The Asyl logic synthesis system incorporates technology mapping for Xilinx 3000 CLBs [Abou90]. The technology mapping phase of Asyl depends upon a reference ordering of the primary input variables that is determined by the logic optimization phase. The Boolean network produced by the logic optimization phase is a lexicographical factorization. If this network is collapsed into a sum-of-products expression, the order of variables within the product terms defines the reference ordering. The technology mapping phase of Asyl consists of two steps. The first step uses slices of the reference ordering to decompose the Boolean network into 4- and 5-input functions, and the second step uses a greedy heuristic to assign these functions to CLBs. Compared to Chortle-crf, on the basis of the results for 3 MCNC networks [Abou90] Asyl requires 7% more CLBs.

3.2.5 The Hydra Technology Mapper

The Hydra technology mapper [Filo91] addresses two-output RAM-based logic blocks such as the Xilinx 3000 series CLBs. The two-phase strategy employed by Hydra to minimize the number of CLBs in the final circuit emphasizes the use of both CLB outputs. The first phase decomposes nodes in the original network to ensure that every node can be implemented by a single CLB and the second phase then finds pairs of functions that can be implemented by two-output CLBs. The first phase creates opportunities for the second phase to pair two functions into a single CLB by selecting decompositions that increase the number of shared inputs among the extracted functions. Using the results for 18 MCNC networks [Filo91], Hydra requires 14% fewer Xilinx 3000 CLBs than Chortle-crf and is 1.5 times faster.

3.2.6 The Xmap Technology Mapper

The Xmap technology mapper [Karp91a] uses two passes to minimize the number of K-input LUTs required to implement a Boolean network and a third pass to produce a circuit of Xilinx 3000 CLBs. The first pass decomposes nodes in the original Boolean network into *if-then-else* DAGs. In an if-then-else DAG, every node functions as a 2 to 1 multiplexer. For values of *K* greater than or equal to 3 every node in the if-then-else DAG can be implemented by a single K-input LUT. The second pass traverses the decomposed network from the primary inputs to primary outputs and greedily marks nodes to be implemented by single-output LUTs. The third pass assigns functions produced by the first two passes to 2-output CLBs. On the basis of the results for 27 MCNC networks [Murg91a], Xmap requires 13% more 5-input LUTs than Chortle-crf to implement the networks, but it is 16 times faster.

3.2.7 The VISMAP Technology Mapper

The VISMAP technology mapper [Woo91a] focuses on the covering problem in LUT technology mapping. It assumes that the original network has been previously decomposed to ensure that every node has in-degree less than or equal to *K*. This ensures that every node can be directly implemented by a K-input LUT.

VISMAP approaches the LUT-mapping problem by labelling every edge in the network as either *visible* or *invisible*. A visible edge interconnects two LUTs in the final circuit and an invisible edge is implemented within a single LUT. The network can be simplified by merging the source node of an invisible edge into the destination node. If the resulting node has in-degree no greater than *K*, then it can still be implemented by a single K-input LUT. The assignment of visibility labels to edges is performed by first dividing the original network into a collection of subgraphs that each contain at most *m* edges. The optimal label assignment for each subgraph is then found using an exhaustive search. The computational cost of the search is controlled by the limit on the number of edges in the subgraph.

VISMAP also includes a greedy heuristic to pair K-input functions into K-input 2-output LUTs. Compared to Chortle-crf, VISMAP requires 8% more 5-input 2-output LUTs to implement 9 MCNC networks [Woo91a], and has similar execution speed.

3.3 Multiplexer Technology Mapping

A multiplexer-based logic block consists primarily of a tree of multiplexers. For example, the Actel Act-1 logic block, illustrated in Figure 3.19, implements the Boolean function

$$\overline{(a+b)}(\overline{c}e+cf)+(a+b)(\overline{d}g+dh).$$

The inputs to the logic block are either the multiplexer select inputs, *a*, *b*, *c*, and *d*, or they are data inputs *e*, *f*, *g*, and *h*. An uncommitted logic block is *personalized* to implement different functions by connecting its inputs either to variables or to constants 0 and 1. For example, the Act-1 logic block can be personalized to implement the function $xy+\overline{x}\,\overline{y}$ by making the input connections $a=x$, $b=0$, $c=y$, $d=y$, $e=1$, $f=0$, $g=0$, and $h=1$.

Multiplexer-based logic blocks can implement a large number of different functions and therefore present difficulties for library-based technology mapping. Examples of technology mappers for multiplexer-based logic blocks include mis-pga [Murg90] [Murg92], Proserpine [Erco91] [Beda92], Amap [Karp91b] and XAmap [Karp91b]. All of these programs map a Boolean network into a circuit of multiplexer-based logic blocks and determine the personalization of every logic block in the circuit. They minimize either the number of logic blocks or the delays in the final circuit. The following sections describe one program, Proserpine, in detail and discuss the

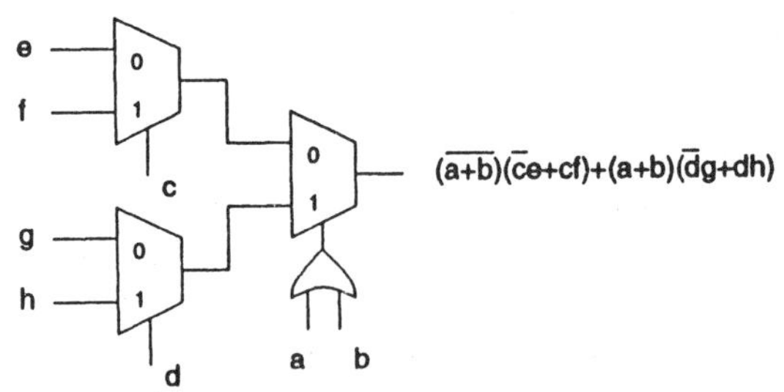

Figure 3.19 - *Act-1 Logic Block.*

major features of the others.

3.3.1 The Proserpine Technology Mapper

The overall approach used by Proserpine is the same as the Ceres [Mail90b] library-based technology mapper. The original network is partitioned into a set of single-output subnetworks and then each subnetwork is mapped separately. The first step in mapping each subnetwork is to decompose it into a network of 2-input functions. This fine-grain decomposition enables the function of nodes in the original network to be split across more than one logic block. However, performing the decomposition before mapping the subnetwork precludes the possibility of optimizing the chosen decomposition.

Similar to DAGON [Keut87], Proserpine uses a dynamic programming traversal to construct a circuit of logic blocks that covers each single-output subnetwork. This traversal relies upon a matching algorithm that determines if a subfunction within the subnetwork, referred to as a *cluster* function, can be implemented by personalizing the multiplexer-based logic block. This matching algorithm models the personalization of a logic block using *stuck-at-0* faults, *stuck-at-1* faults, and *bridging* faults. It uses Binary Decision Diagrams (BDDs) [Brya86] to represent the cluster function and the logic block, and uses subgraph isomorphism to detect a match.

3.3.1.1 Binary Decision Diagrams

A BDD is a two terminal DAG with a single root node. The terminal nodes represent the values 0 and 1, while non-terminal nodes represent Boolean functions. The function associated with the root node specifies the function represented by the entire BDD. Each non-terminal node has an associated variable and two outgoing edges, labelled 0 and 1. The function represented by the non-terminal node is specified by its cofactors with respect to its associated Boolean variable. The subDAGs terminating the two outgoing edges specify these cofactors. If the sequence of variables along any path in the BDD is restricted to a given precedence order and if no isomorphic subgraphs exist within the BDD, then the result is a canonical form known as a *reduced ordered* BDD. For example, Figure 3.20a illustrates the multiplexer-based logic block $\bar{a}(\bar{b}c+bd)+a(\bar{e}f+eg)$ and Figure 3.20b illustrates the BDD representing this logic block corresponding to the input variable ordering (a, b, c, d, e, f, g). In the remainder of this section the term BDD will refer to a reduced ordered BDD. Note that the structure of the BDD depends on the input variable precedence order.

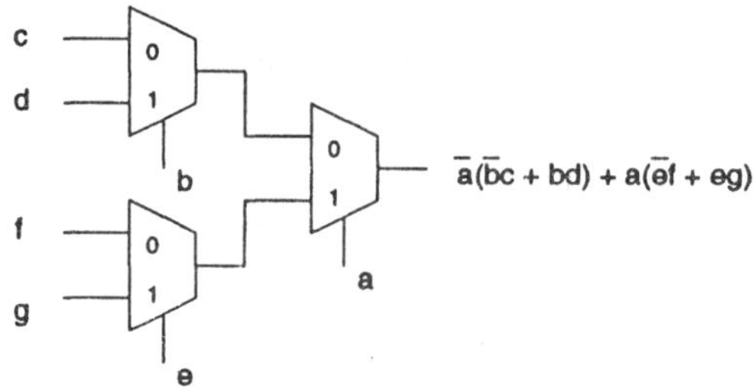

a) multiplexer-based logic block

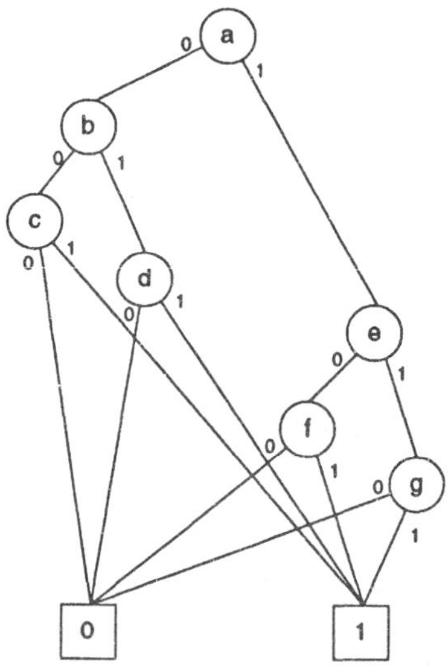

b) Binary Decision Diagram

Figure 3.20 - *A multiplexer Logic Block and Its BDD.*

The personalization of a logic block to match a cluster function is defined by a set of stuck-at faults, a set of bridging faults, and an input variable assignment. The matching algorithm consists of two stages. The first stage considers only stuck-at faults and the second stage considers bridging faults in addition to stuck-at faults.

3.3.1.2 Matching with Stuck-At Faults

The first stage of the matching algorithm considers a simplified matching problem in which only stuck-at-0 and stuck-at-1 faults are used to personalize the logic block. The BDD representing the cluster function is compared to subgraphs in the BDD representing the logic module. If a subgraph is isomorphic to the cluster function BDD, then it represents the same Boolean function. In this case, the logic block can be personalized to implement the Boolean function by the appropriate set of stuck-at faults. The required set of stuck-at faults is specified by the path leading from the root of the logic block BDD to the subgraph, and the input assignment is specified by the correspondence between nodes in the cluster function BDD and the nodes in the subgraph. For example, in Figure 3.21 the BDD for the cluster function $\overline{x}y+xz$ is isomorphic to a subgraph in the BDD for the logic block $\overline{a}(\overline{b}c+bd)+a(\overline{e}f+eg)$. Therefore, the logic block can be personalized to implement this cluster function using the stuck-at fault $a=1$ and the input assignment $e=x$, $f=y$, and $g=z$. Note that the assignment of inputs b, c, and d does not matter.

The existence of a subgraph isomorphic to the cluster function BDD depends on the structure of the logic block BDD, and therefore depends on the input ordering used to construct the logic block BDD. For example, in Figure 3.21 the input ordering (a, b, c, d, e, f, g) is used to construct the logic block BDD and there exists a subgraph that is isomorphic to the cluster function BDD. However, if the input ordering (a, f, e, g, d, b, c) is used to

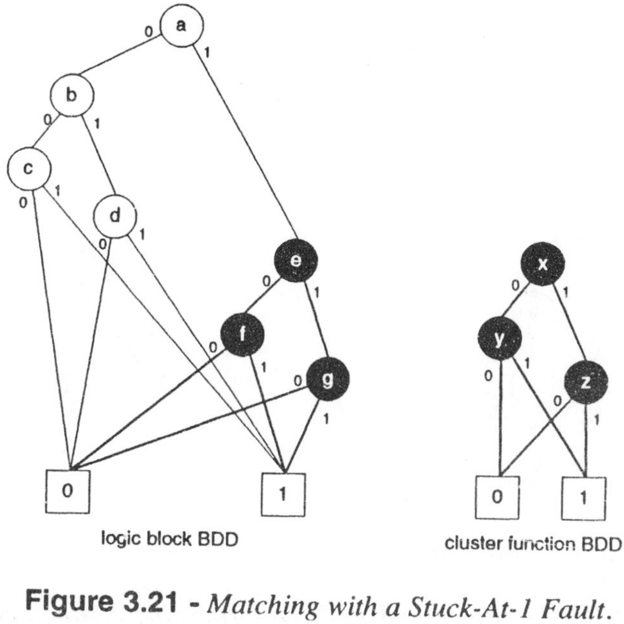

Figure 3.21 - *Matching with a Stuck-At-1 Fault.*

construct the logic block BDD, as illustrated in Figure 3.22, then there is no isomorphic subgraph.

To ensure that a match is found, regardless of the input ordering used to construct the logic block BDD, the first stage of the matching algorithm, outlined as pseudo-code in Figure 3.23, considers all possible input orderings for the logic block BDD and searches each logic block BDD for a subgraph that is isomorphic to the cluster function BDD. The size of the search is reduced by restricting it to subgraphs of the same height as the cluster function BDD.

Many of the subgraphs within the logic block BDDs corresponding to different input orderings will be isomorphic to one another. Only one of these subgraphs needs to be considered in the search for a subgraph isomorphic to the cluster function BDD. Proserpine reduces the size of the search for an isomorphic subgraph by assembling the logic block BDDs for all possible input orderings into one common structure referred to as a Generalized Binary Decision Diagram (GBDD). Within the GBDD there are no subgraphs that are isomorphic to each other. The two loops in Figure 3.23

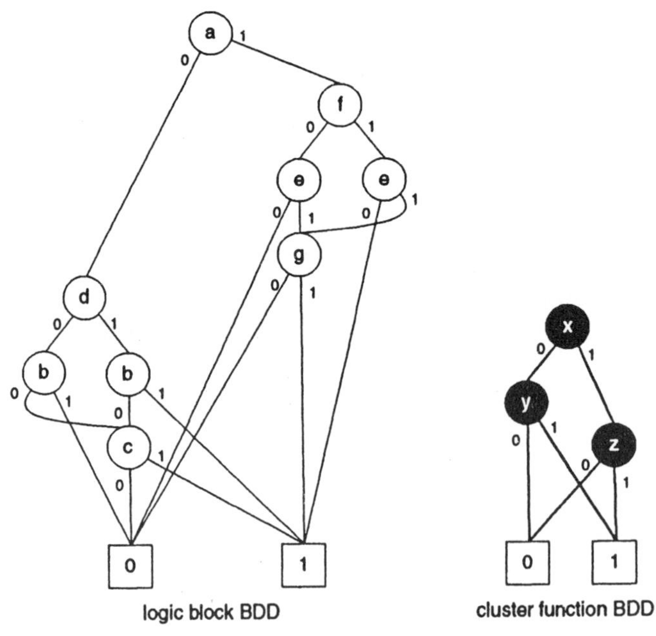

Figure 3.22 - *Logic Block BDD Ordering where Match Fails.*

```
StuckMatch (cluster, module) {
    /* match using stuck-at faults */

    clusterBDD ← buildBDD (cluster, DefaultOrder)
    clusterHeight ← height of clusterBDD

    for all module inputOrders {
        moduleBDD ← buildBDD (module, inputOrder)

        for all subGraphs of clusterHeight in moduleBDD {
            if isomorphic (clusterBDD, subGraph) {
                record stuckAtFault
                record inputOrder
                return (Match)
            }
        }
    }
    return (NoMatch)
}
```

Figure 3.23 - *Pseudo-code for Stuck-At-Fault Matching.*

are collapsed into one loop that considers all subgraphs within the GBDD.

3.3.1.3 Matching with Bridging Faults

If the first stage of the matching algorithm fails to find a set of stuck-at faults that personalizes the logic block to the cluster function, then the second stage considers personalizations that require bridging faults. The presence of a bridging fault modifies the logic block BDD and can create a subgraph that is isomorphic to the cluster function BDD. Consider the logic block BDD shown in Figure 3.24a. This logic block implements the functions $F = \bar{a}(\bar{b}F_{\bar{a}\bar{b}} + bF_{\bar{a}b}) + a(\bar{b}F_{a\bar{b}} + bF_{ab})$, where F_{ab}, $F_{a\bar{b}}$, $F_{\bar{a}b}$, and $F_{\bar{a}\bar{b}}$, are the cofactors of F with respect to the variables a and b.

If there is a bridging fault between the variables a and b, then only the cofactors F_{ab} and $F_{\bar{a}\bar{b}}$ remain possible and the logic block BDD can be simplified to the BDD shown in Figure 3.24b. The matching algorithm can then search for a subgraph in the modified logic block BDD that is isomorphic to the cluster function BDD. A similar modification to the logic block BDD can be made for any fault that bridges one or more sets of adjacent variables in the BDD. An arbitrary bridging fault can be described by

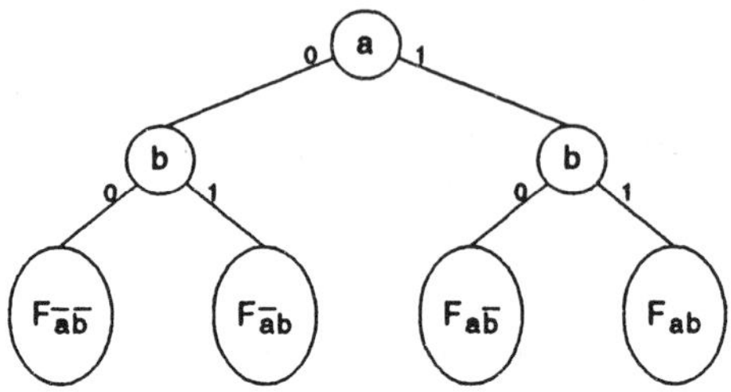

a) Logic block BDD

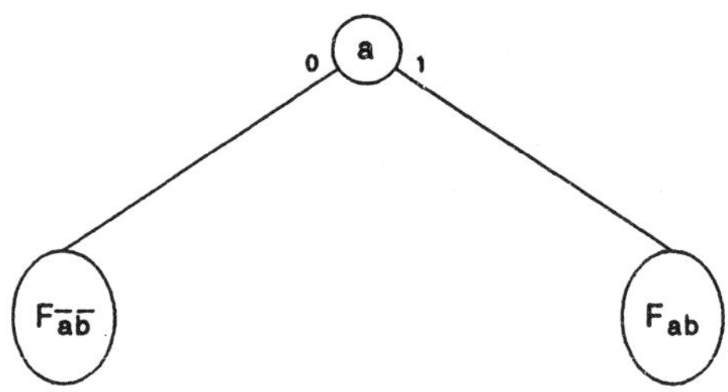

b) BDD with bridge fault $a = b$

Figure 3.24 - *Simplifying the Logic Block BDD with a Bridge Fault.*

finding an input ordering where the bridged variables are adjacent and then modifying the corresponding logic block BDD.

The second stage of the matching algorithm, outlined as pseudo-code in Figure 3.25 considers all possible bridging sets and for each of these bridging sets considers all possible input orderings. Each bridging set specifies the variable positions to be bridged. The actual variables that are bridged depend on the variable ordering. For each bridging set and input order, the algorithm constructs the corresponding logic block BDD and searches for subgraphs that match the cluster function BDD. The GBDD that represents the logic block BDDs for all possible input orderings can be used to reduce the size of this search. In this case, the bridge set modifies the entire GBDD and the two inner loops of Figure 3.25 are collapsed into one loop that considers all subgraphs of the modified GBDD.

3.3.1.4 Matching with One Bridge Fault

The bridge-fault matching algorithm is much more computationally expensive than the stuck-at-fault matching algorithm, yet in experimental results, it only slightly decreased the number of logic blocks in the final

```
BridgeMatch (cluster, module) {
    /* match using bridging faults */

    clusterBDD ← buildBDD (cluster, DefaultOrder)
    clusterHeight ← height of cluster

    for all bridgeSets of adjacent variables {

        for all module inputOrders {
            moduleBDD ← buildBDD (module, inputOrder)
            bridgeBDD ← bridge (moduleBDD, bridgeSet)

            for all subGraphs of clusterHeight in bridgeBDD {
                if isomorphic (clusterBDD, subGraph) {
                    record stuckAtFault
                    record bridgeSet
                    record inputOrder
                    return (Match)
                }
            }
        }
    }
    return (NoMatch)
}
```

Figure 3.25 - *Pseudo-code for Bridge-Fault Matching.*

circuit. It was observed that most of the bridge faults found consisted of one bridge of two inputs. To reduce the computational cost of finding these bridge faults an alternative bridge-fault matching algorithm is introduced. This simplified algorithm only searches for bridge faults that consist of one bridge of two inputs.

The key to the one-bridge matching algorithm is the observation that one bridge of two inputs can be expressed as a pair of stuck-at faults. Consider the subgraph of the logic block BDD, corresponding to input ordering (f, g, d, e, b, a, c), and the cluster function BDD shown in Figure 3.26. If the logic block matches the cluster function when the variable x is assigned to inputs e and b, then the subgraph of the cluster function specified by the stuck-at fault $x=0$ must be isomorphic to the subgraph specified by the stuck-at fault $e=0$ and $b=0$, as illustrated in Figure 3.26a. Similarly, the

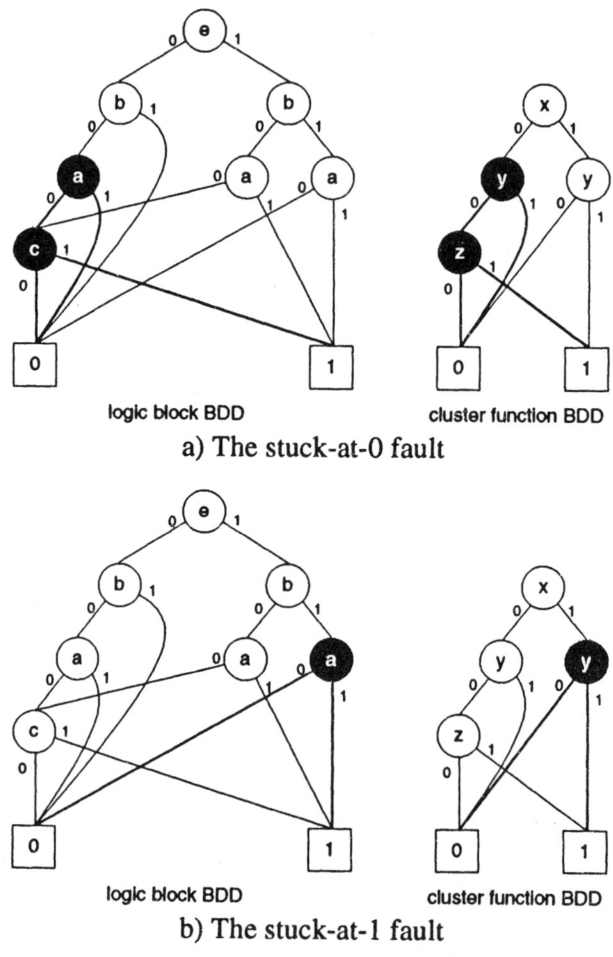

Figure 3.26 - *Matching a Bridge Fault as a Pair of Stuck-At Faults.*

subgraph of the cluster function specified by the stuck-at fault $x=1$ must be isomorphic to the subgraph specified by the stuck-at fault $e=1$ and $b=1$, as illustrated in Figure 3.26b. This match associates the bridged inputs, e and b, to the first cluster variable x. Bridge faults that use a different cluster variable can be found by considering cluster function BDDs with different variable orderings.

The one-bridge matching algorithm, outlined as pseudo-code in Figure 3.27, considers each of the cluster function variables in turn, and for each

```
OneBridgeMatch (cluster, module) {
    /* match using one bridge fault */

    variableOrder ← DefaultOrder
    for all variables of cluster {
        place variable at head of variableOrder
        clusterBDD ← buildBDD (cluster, variableOrder)
        clusterHeight ← height of clusterBDD

        leftCluster ← first variable of clusterBDD stuck-at-0
        rightCluster ← first variable of clusterBDD stuck-at-1

        for all module inputOrders {
            moduleBDD ← buildBDD (module, inputOrder)

            for all subGraphs of clusterHeight + 1 in moduleBDD {
                leftModule ← first two inputs of subGraph stuck-at-0
                rightModule ← first two inputs of subGraph stuck-at-1

                if isomorphic (leftCluster, leftModule) &
                    isomorphic (rightCluster, rightModule) {
                    bridgeSet ← first two inputs of subGraph
                    record stuckAtFault
                    record bridgeSet
                    record inputOrder
                    record variableOrder
                    return (Match)
                }
            }
        }
    }
    return (NoMatch)
}
```

Figure 3.27 - *Pseudo-code for One-Bridge-Fault Matching.*

variable constructs a cluster function BDD with the variable as the first variable in the BDD ordering. For each of these cluster function BDDs, the algorithm considers all possible input orderings for the logic block. For each input ordering, the algorithm constructs the corresponding logic block BDD and searches for subgraphs of this BDD where bridging the first two variables of the subgraph to the first variable of the cluster function BDD results in the required pair of stuck-at-fault matches. Note that only subgraphs of height one greater than the height of the cluster BDD need to be searched because the first two inputs of the subgraph will be bridged together. The size of the search can be reduced by using the GBDD to represent the logic block BDDs for all possible input orderings and collapsing the inner two loops of Figure 3.27 into one loop that considers all subgraphs of the GBDD.

3.3.1.5 Proserpine Performance

This section presents the results of using Proserpine to map 16 MCNC networks into circuits of Act-1 and Act-2 logic blocks. Column 1 of Table 3.3 lists the names of the networks, and columns 2,3 and 4 record the number of Act-1 logic blocks in the circuits constructed using stuck-at, bridging, and one-bridge matching. Columns 5,6 and 7 record the number of Act-2 logic blocks in the circuits constructed using only stuck-at, stuck-at with full bridging, and stuck-at with one-bridge matching.

network	Act-1			Act-2		
	stuck	bridge	one	stuck	bridge	one
5xp1	54	50	50	48	46	48
C1908	280	206	207	209	204	206
C499	274	170	170	170	170	170
C5315	912	796	796	729	723	724
apex6	411	383	383	295	288	292
apex7	122	114	114	108	104	107
bw	67	63	63	64	61	64
clip	74	68	68	62	60	62
des	1783	1673	1673	1404	1384	1404
duke2	178	175	175	164	162	162
f51m	65	59	59	52	50	52
misex1	24	24	24	23	23	23
misex2	46	42	42	40	39	40
rd84	75	65	65	63	61	63
vg2	46	45	45	41	41	41

Table 3.1 - *Proserpine Results, Act-1 and Act-2 logic blocks.*

Compared to only stuck-at matching, stuck-at with bridging matching reduces the number of Act-1 logic blocks by 11%, and the number of Act-2 logic blocks by 2%. One-bridge matching is nearly as effective as bridging matching for the Act-1 logic block, but, it is less effective for the Act-2 logic block. Note also that fewer Act-2 than Act-1 logic blocks are required to implement the networks. This indicates the increased functionality of the Act-2 logic block.

3.3.2 Multiplexer Technology Mapping in mis-pga

The objective of the mis-pga [Murg90] technology mapper is to minimize the number of Actel Act-1 logic blocks required to implement a Boolean network. This program converts each node in the original network into a BDD and then uses dynamic programming to cover the resulting graph with a set of 8 pattern graphs that represent the Act-1 logic block. A final iterative improvement phase performs local transformations on the circuit to improve the final result.

A later version of mis-pga [Murg92] maps the Boolean network into a circuit of either Actel Act-1 or Act-2 logic blocks. Each node of the network is converted into an if-then-else DAG, using recursive cofactoring. The advantage of if-then-else DAGs over BDDs is that they avoid the duplication of cubes. Dynamic programming is used to cover the if-then-else DAG with a small set of pattern graphs that represent the logic block. The algorithm uses different matching algorithms for the Act-1 and Act-2 logic blocks to determine if a subgraph of the if-the-else DAG can be implemented by a logic block. These specialized matching algorithms take advantage of the precise structures of the Act-1 and Act-2 logic blocks. On the basis of results for 17 MCNC networks [Murg92], mis-pga is faster than Proserpine, and requires 25% fewer Act-l logic blocks to implement the networks.

3.3.3 The Amap and XAmap Technology Mappers

The Amap and XAmap technology mappers [Karp91b] attempt to minimize the number of Actel Act-1 logic blocks required to implement a Boolean network. Amap begins by decomposing nodes in the original Boolean network into *if-then-else* DAGs. It then proceeds from the primary outputs to the primary inputs using greedy heuristics to cover subDAGs of the if-then-else DAG with the Act-1 logic block. Using the results for 17 MCNC networks [Murg92], Amap requires 12% fewer Act-1 logic blocks and is much faster than Proserpine.

XAmap is based on the observation that the Act-1 logic block can implement 213 of the 256 possible 3-input functions. XAmap begins by mapping the Boolean network into a circuit of 3-input LUTs, using Xmap. Any LUTs that implement one of the 213 functions can be implemented by a single Act-1 logic block. The remaining LUTs can be implemented by one Act-1 logic block provided that one of the three inputs is available in both positive and negative polarities. If none of the three inputs is available in both polarities, then an extra logic block is used to invert one of the signals.

3.4 Final Remarks

Conventional library-based technology mapping is inappropriate for FPGAs because the complex logic blocks used in FPGAs can each implement a large number of different functions. This chapter has described technology mappers that deal specifically with LUT-based and multiplexer-based logic blocks. The key features of these programs are matching algorithms that determine if a subfunction of the network being mapped can be implemented by a single logic block. These matching algorithms avoid the large library of individual functions required to represent the logic block by using the structure of the logic block itself to represent the entire set of functions. In addition, this simplified matching makes it possible to improve the final circuit by simultaneously solving the decomposition and matching problems.

CHAPTER 4
Logic Block Architecture

Chapter 2 described many of the commercial FPGA architectures, but provided little comment on the relative merits of each. This chapter focuses on the design of one aspect of FPGA architecture, namely the architecture of the logic blocks. We discuss the effect of the logic block design on both the total chip area needed in an FPGA to implement a given amount of logic, and the speed performance of an FPGA. The results of several studies on this topic are compared and contrasted, using a consistent notation and style of presentation.

An important characteristic of a logic block is its *functionality*, which is defined as the number of different boolean logic functions that the block can implement. For example, a two-input NAND gate can implement five different functions: the basic function $f=\overline{ab}$, as well as $f=\overline{a}$ and $f=\overline{b}$, and the constants 0 and 1, if the inputs are set appropriately. In contrast, a three-input lookup table can implement any function of its three inputs, providing a much greater functionality of $2^{2^3}=256$.

There are many different architectural choices that could be made for a logic block, as apparent from the examples in Chapter 2. Different blocks are likely to have different amounts of functionality, and varying costs in terms of chip area and delay. Also, the functionality of the logic block will affect the amount of routing resources that are needed in the FPGA. As noted earlier, FPGA routing resources are expensive in terms of area and delay because the programmable switches take up significant area and have appreciable resistance and capacitance. This chapter will show that it is this latter issue that dominates the logic block architecture tradeoffs.

The chapter presents some recent research results on the best choice of logic block functionality. We assume that an FPGA consists of an array of identical (homogeneous) blocks. The chapter is divided into two parts: the first deals with the effect of logic block functionality on FPGA area, and deals with lookup table-based logic blocks. The second part covers the effect of functionality on speed performance, and includes several different types of blocks.

4.1 Logic Block Functionality versus Area-Efficiency

The functionality of the logic block has a major effect on the amount of area required to implement circuits in FPGAs. As functionality increases, the number of blocks needed to implement a circuit will decrease, but the area per block will increase, because higher functionality requires more logic. It follows, then, that there is an amount of functionality in the logic block that achieves a minimum total area for the logic blocks themselves.

The total chip area needed for an FPGA consists of the logic block area plus the routing area. Since routing area typically takes from 70 to 90 percent of the total area, the effect of logic block functionality on the routing area can be very important. Functionality affects the routing area in a number of ways: as functionality increases, the number of pins per block will likely increase, the number of connections between logic blocks will decrease and, because there will be fewer blocks, the distance that each connection travels will decrease. Depending on the relative effects on each of these factors, the total routing area will either go up or down.

For example, consider the implementation of the logic function $f = abd + bc\bar{d} + \bar{a}\bar{b}\bar{c}$ in logic blocks with different functionalities. Figure 4.1 illustrates how this function could be implemented with 2-input, 3-input, or 4-input lookup tables. We will refer to a lookup table as a LUT, and a K-input lookup table as a K-LUT. As shown, the 2-LUT implementation requires eight logic blocks, while the 3-LUT needs only four blocks. As an area measure, consider the number of memory bits required for each implementation. A 2-LUT requires half as many memory bits as a 3-LUT (recall that the number of bits in a K-LUT is 2^K), but twice as many 2-LUTs are required. Hence, the total block area for both the 2-LUT and 3-LUT cases is the same. The 4-LUT case requires only half the number of memory bits compared to the other two, because the function can be implemented in just one block. However, any one of the three alternatives may result in the lowest total chip area, depending on the amount of routing resources that each one implies, as discussed below.

The routing area for each implementation can change dramatically. As Figure 4.1 shows, the number of connections to logic block inputs and outputs for the 2-LUT, 3-LUT and 4-LUT implementations are 17, 13 and 5, respectively. Depending on the length of the wires required to implement the connections in each case, any one of the three implementations may have the smallest routing area. For example, if a wire in the 2-LUT case is shorter, then it may be better to have 17 of those wires compared to only 13 of the longer wires for the 3-LUT case. The salient point of this discussion is that the effects of the functionality of the logic block on total chip area are

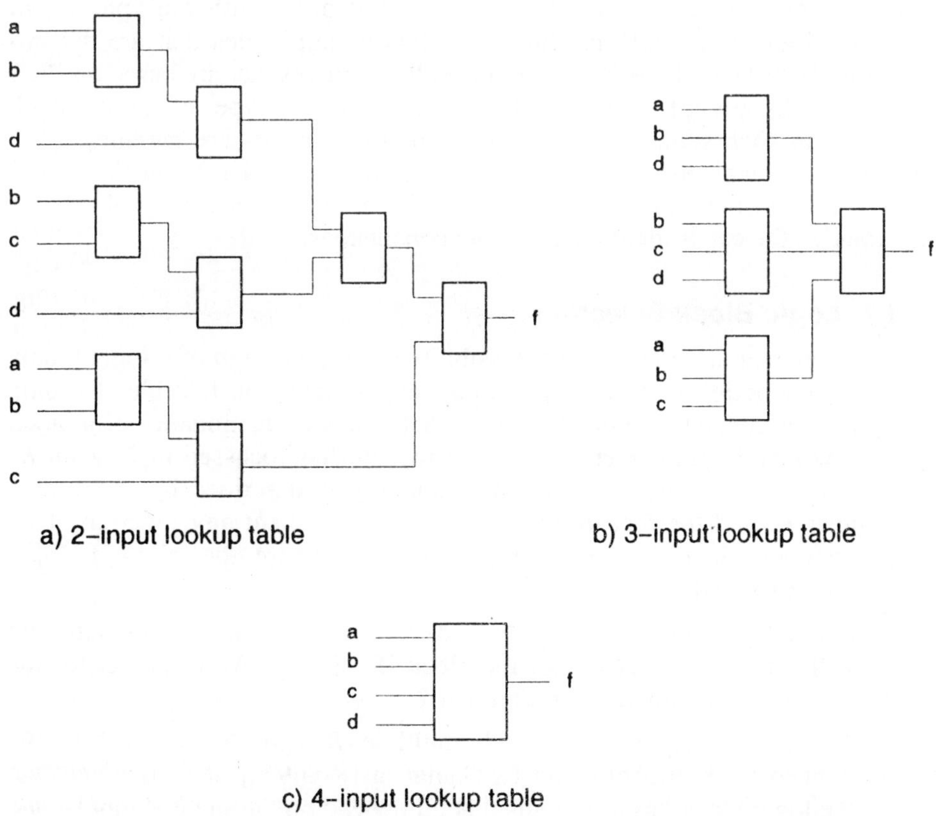

Figure 4.1 - *Three Implementations of* $f = abd + bc\bar{d} + \bar{a}b\bar{c}$.

complex, and involve both the area due to the logic block itself and the routing resources that interconnect the blocks.

The goal of the experimental studies presented in this chapter is to answer the question: "What level of functionality gives the lowest total chip area for an FPGA?" The above example and discussion serves to motivate the experimental approach that has been used in all of the research addressing this question. In this approach, benchmark circuits are "implemented" using a CAD system that can handle a range of different logic block architectures. To measure the results, the studies use models that account for both logic block area and routing area.

The following sections summarize four recent studies of the area effects of logic block architectures. The first study examines single-output lookup tables [Rose89] [Rose90c], the second deals with multiple-output lookup tables [Koul92a], the third considers lookup tables that are decomposable [Hill91], and the fourth examines logic blocks that are based on PLA structures [Koul92b]. Section 4.1.1 describes the type of logic block assumed in each study and shows how the experiments are parameterized. The experimental procedures are described in Section 4.1.2 and the model that is used to measure area is given in Section 4.1.3. In Section 4.1.4, we summarize the experimental results and conclusions.

4.1.1 Logic Block Selection

There is a large number of possibilities for the design of a logic block, so some restriction is necessary to make experimentation feasible. To gain insight, it is important to be able to characterize the functionality of a block by a few simple parameters. Research to date has focussed mainly on K-input lookup tables because their functionality is described by the simple parameter K, and the fact that the block can implement any function of its inputs is convenient. In the four studies described here, the following logic blocks are explored:

1. A single-output block, with one K-input lookup table, both with and without a D Flip as part of the block [Rose90c]. An example, for the case $K=4$, is illustrated in Figure 4.2.
2. A multiple-output block, with multiple K-input lookup tables. As defined by Kouloheris and El Gamal in [Koul92a], it is assumed that the logic block has M outputs that all depend on the same K inputs, and each output is generated by a separate lookup table.
3. A multiple-output block, with *decomposable* lookup tables, as defined by Hill and Woo in [Hill91]. For this investigation, it is assumed that

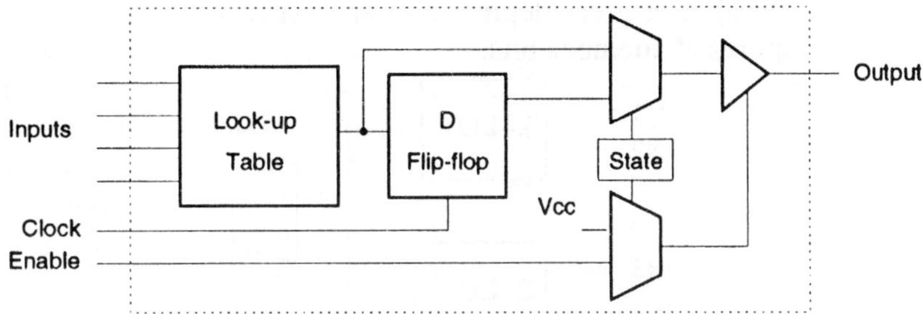

Figure 4.2 - *Single Output 4-LUT Logic Block, with a D flip-flop.*

the logic block contains a total of M K-input lookup tables. However, the block is decomposable, meaning that it can be viewed as a set of M K-LUTs, or $M-2$ $(K+1)$-LUTs, $M-4$ $(K+2)$-LUTs, and so on. This is based on the fact that smaller lookup tables can be combined to implement larger lookup tables. An example is given in Figure 4.3, which shows a logic block that can be viewed as two 2-LUTs or as one 3-LUT. To implement two 2-LUTs, both of the outputs, $OUT1$ and $OUT2$, would be used. Note that $IN5$ would be forced to 0 in this case, so that the multiplexer would select the lower 2-LUT. The other alternative is to view the block as a 3-LUT having a single output, $OUT2$. In this case, inputs $IN1$ and $IN3$, as well as $IN2$ and $IN4$, should be connected together, while $IN5$ serves as the third input to the 3-LUT. The top 2-LUT then implements one half of a truth-table, the bottom 2-LUT implements the other, and the multiplexer selects whichever half is required.

For large values of M and K, the number of inputs to the logic block may be prohibitive. Since this has a major impact on the total area required for routing resources, Hill and Woo examine the reduction of the number of inputs by considering the sharing of inputs between pairs of LUTs. In this scheme, the two LUTs that feed a common multiplexer are considered as a pair and are forced to share Z of their inputs in common. This allows larger values of M and K to be used without overly increasing the total number of inputs to the block. Z is mentioned here because it is an interesting architectural parameter, but we do not show experimental results for Z in this book.

4. A PLA-based block, having K inputs, N product terms, and M outputs, as described in [Koul92a] and [Koul92b]. The

motivation for this block is that larger lookup tables are under-utilized to a great degree, and are expensive because a K-LUT requires 2^K memory bits.

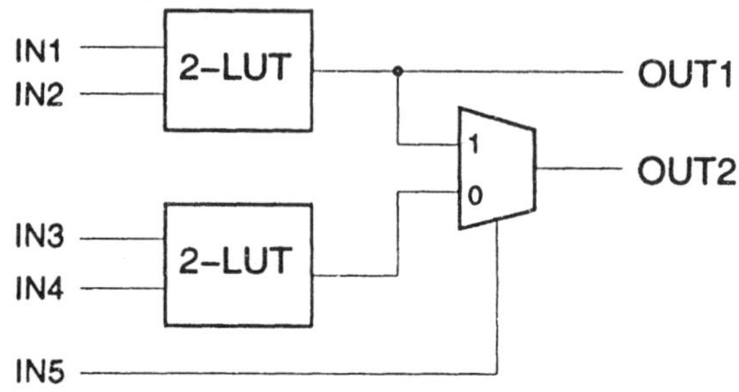

Figure 4.3 - *A Decomposable LUT Block.*

4.1.2 Experimental Procedure

This section describes the experimental framework that has been used in all four of the research studies that are discussed in this chapter. The input to each procedure is a logic circuit, a functional description of the logic block, and the FPGA's programming technology, which was discussed in detail in Chapter 2. The output of the procedure is the total chip area required to implement the circuit. The following steps are applied for each circuit, programming technology, and logic block:

1. Perform the technology mapping of the circuit into the type of logic blocks assumed. This determines the total number of logic blocks required for the circuit. Each of the four studies discussed here uses a different technology mapping program for this task. The interested reader can refer to the original publications for more details. The result of this step is a new circuit that consists of only the available type of logic block.

2. Perform the placement of the logic blocks. This step assigns each logic block to a specific location in the FPGA.

3. Perform the global routing of the circuit. Global routing selects the paths through the channels that each connection should take. This step determines the number of tracks per routing channel that are required for the circuit. The number of tracks is called W.

In a real FPGA, the number of logic blocks and tracks per channel would be fixed, but in the above procedure, both of these are measured as the outputs of the experimental procedure. The following section shows how these two measurements can be used to calculate the total area required in the

FPGA.

Note that the experiments in [Hill91] did not proceed to the placement and routing level, but rather calculated the area measures based on the total number of logic blocks and the number of inputs to a block only.

To estimate the routing area, it is necessary to make assumptions about the interconnection architecture. In [Rose90c], the symmetrical architecture illustrated in Figure 4.4 is used. It is a regular array of identical logic blocks, separated by horizontal and vertical routing channels. The number of tracks in all of the routing channels, W, is the same. In [Koul92a] and [Koul92b], a row-based architecture similar to that in Actel FPGAs is assumed with W tracks per row. No assumption is necessary about routing structures in [Hill91] because circuits are not synthesized to the level of detail of placement and routing.

4.1.3 Logic Block Area and Routing Model

A crucial part of each experimental approach is the modelling of logic and routing area as a function of the type of logic block used and the definitions of the routing structures. This section unifies the models that appear in [Rose90c], [Koul92a] and [Hill91].

Recall from Chapter 2 that the *programming technology* for an FPGA refers to the technology used to implement its programmable resources. In Chapter 2, the programming technology was defined specifically in the context of implementing routing switches, but it could also implement

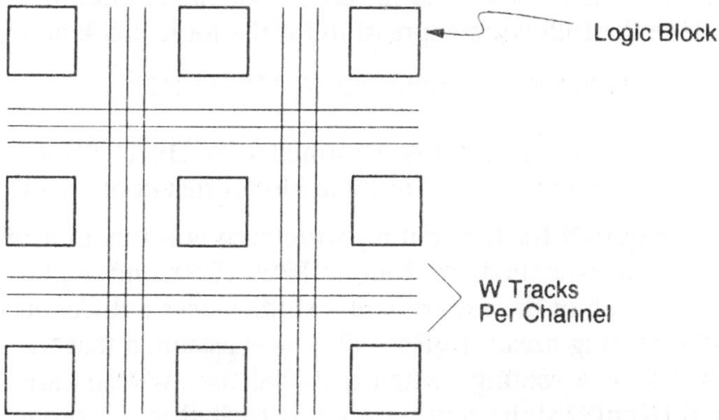

Figure 4.4 - *Interconnection Model of the FPGA.*

programmable resources within the logic blocks, such as the memory bits of a lookup table. In this chapter, it is assumed that the programming technology determines how both the routing switches and the lookup table memory bits are implemented. For this reason, there are two area parameters that are dependent on the programming technologies: BA and RP. *BA* stands for Bit Area, and refers to the area required to implement each memory bit of a lookup table. *RP* corresponds to the Routing Pitch, which is determined by the size of a routing switch, as explained shortly.

As an example, in a static RAM-based FPGA *BA* is the area of a Static RAM bit (roughly $400\mu m^2$ in 1.2 μm CMOS), and for an antifuse-based FPGA it is the size of the antifuse and the associated programming transistors (about $40\mu m^2$). Similarly, for EPROM-based FPGAs, *BA* is the size of an EPROM transistor and associated programming circuitry.

The following section describes the logic and routing model used to calculate the area for lookup table-based FPGAs. The subsequent section describes a model for PLA-based FPGAs.

4.1.3.1 Lookup Table-Based FPGAs

The area of a logic block of the form shown in Figure 4.2 is composed of two parts: the area needed for a *K*-LUT, which is a function of *K*, and a fixed area for all other circuitry. The variable area for a *K*-LUT is proportional to the 2^K memory bits needed in its implementation. If there are *M* *K*-LUTs, then there must be $M \times 2^K$ memory bits. The fixed area is called *FA*, and includes circuitry required to access the *K*-LUT(s), the area required by the D flip-flop (if it is present) and all other associated circuitry. Using FA and BA, we have the following expression for the logic block area:

$$Logic\ Block\ Area = FA + (M \times BA \times 2^K) \qquad 4.1$$

In a 1.25μm CMOS technology, FA is estimated as $2100\mu m^2$ for logic blocks without a D flip-flop and $5100\mu m^2$ for logic blocks that contain a D flip-flop.

The area required for the routing structures is a function of the space needed between tracks, called the Routing Pitch (RP), and the dimensions of the logic block. Each of the experimental studies uses a different expression to calculate the routing area. In [Rose90c], it is assumed that tracks must be spaced the width of a routing switch and that the switches are square. In [Koul92a] and [Koul92b] the routing pitch is treated in a more general way, and is varied independently as an experimental parameter. In [Rose90c], routing area per block is calculated as follows: given that the area of the logic block is known, it is assumed that the block is square, so that the length of one side of the block, *CL*, is given by $CL = \sqrt{Logic\ Block\ Area}$. From the illustration in Figure 4.5 the routing area per block can be calculated as:

$$\textit{Routing Area Per Block [Rose90c]} = 2\,(CL \times W \times RP) + (W \times RP)^2 \qquad 4.2$$

In [Koul92a] the routing area is calculated differently because it assumes a row-based FPGA. In that case, if each channel has *W* wires, then the routing area per block is given by:

$$\textit{Routing Area Per Block [Koul92a]} = CL \times W \times RP \qquad 4.3$$

Recall that [Hill91] does not perform the placement and routing steps. Their model accounts for routing area by simply counting the number of pins on the logic block, and assuming that each one requires a fixed amount of interconnection area. Other research has shown that routing area does correlate well with the number of pins on a logic block, so this is a reasonable approximation. The expression used to calculate logic block area is equivalent to Equation 4.1, with $M = 1$, but the routing area is given by

$$\textit{Routing Area Per Block [Hill91]} = C \times BA \left[M \times K - \left(\frac{M}{2} \times Z\right) + (2M - 1) \right] \qquad 4.4$$

Here, C is a constant for routing area per logic block pin, and the remainder of the expression simply counts the numbers of input and output pins on the block, including those for multiplexers such as the one that was shown in Figure 4.3. Notice that there are fewer pins when pairs of *K*-LUTs share Z inputs. According to [Hill91], a reasonable value for C is equivalent to the

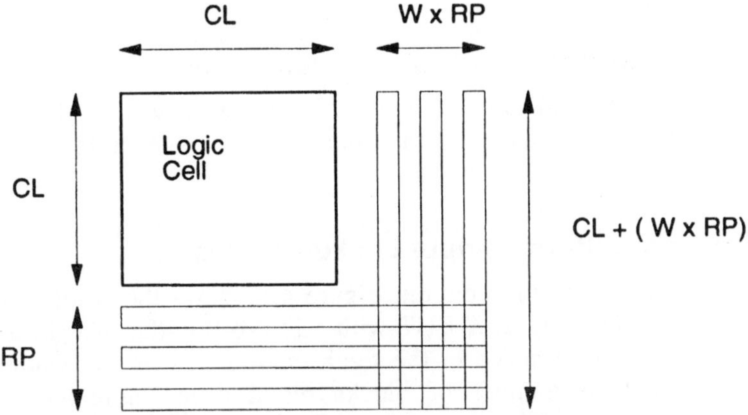

Figure 4.5 - *Routing Area Model for [Rose90c].*

area for eight LUT memory bits.

The total area for a circuit implemented with a particular logic block is then given by:

$$Total\ Area = N_{block}\,(Logic\ Block\ Area + Routing\ Area\ Per\ Block)\ ,$$

where N_{block} is the number of logic blocks needed to implement the circuit, which is determined by the technology mapping.

4.1.3.2 PLA-Based FPGAs

As mentioned in Section 4.1.1, [Koul92a] and [Koul92b] study logic blocks based on K-input, M-output PLAs that can OR together a maximum of N product terms of the K inputs. The PLA model used is a pseudo-static NMOS PLA, and assumes that the programming elements are square. The logic block area is calculated differently than for the lookup tables. The following relation calculates the width of the PLA-based logic block:

$$C_w = \max(18, \sqrt{BA}) \times K + \max(10, \sqrt{BA}) \times M + 98\ .$$

The block height is calculated as

$$C_h = \max(10, \sqrt{BA}) \times N + 136$$

and the total block area is then

$$Logic\ Block\ Area\ (PLA)\ [Koul92a] = C_w \times C_h + FA \times M\ ,$$

where FA is the fixed area per output. The routing area is calculated using Equation 4.3.

4.1.4 Experimental Results and Conclusions

The experimental procedure described above was performed on a variety of circuits in each of the four studies. The following sections presents the results.

4.1.4.1 Single Output, K-Input Lookup Tables

For a typical example circuit, Figure 4.6 shows the results of experiments for the single output K-LUT with a D flip-flop that was illustrated in Figure 4.2. The solid curve in the figure shows that as the value of K is increased, the total number of blocks required to implement a circuit decreases. This makes intuitive sense since larger lookup tables can implement more logic. The dotted curve in the figure shows the area required for a single logic block, which increases exponentially with K, as defined by

Equation 4.1. Typical values are chosen for the BA and FA parameters to generate these results. The total area required for the logic blocks is given by the product of the two curves in Figure 4.6. However, this is only a part of the chip area needed in the FPGA, since it does not account for the routing structures.

Figure 4.7 shows the effect of *K* on the area needed for routing structures, on a per logic block basis. The solid curve is the same as for Figure 4.6 and the dotted curve was calculated using Equation 4.2. For these results, typical values were chosen for *CL* and *RP*, and *W* was determined from the experimental procedure. The reason that the dotted curve increases for higher values of *K* is that *W* increases as *K* does. Again this makes intuitive sense because a logic block with more inputs will require more interconnections. The total area in the FPGA needed for the routing structures is given by the product of the two curves.

Figures 4.6 and 4.7 show the individual area requirements of the logic blocks and the routing structures. Combining the two results yields the total chip area required for the FPGA. Figure 4.8 illustrates this by showing a summary of results for several example circuits. The figure shows a family of curves, where each one corresponds to a different value of BA. Each curve in the figure represents an average over a set of 12 circuits. For each circuit, the total area is normalized to the smallest area that was achievable over all values of K.

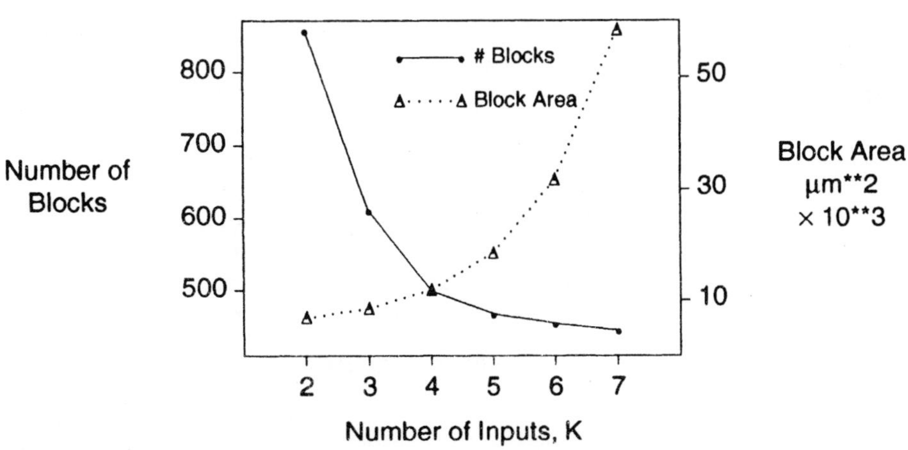

Figure 4.6 - *No. of Blocks and Block Area, for one Circuit.*

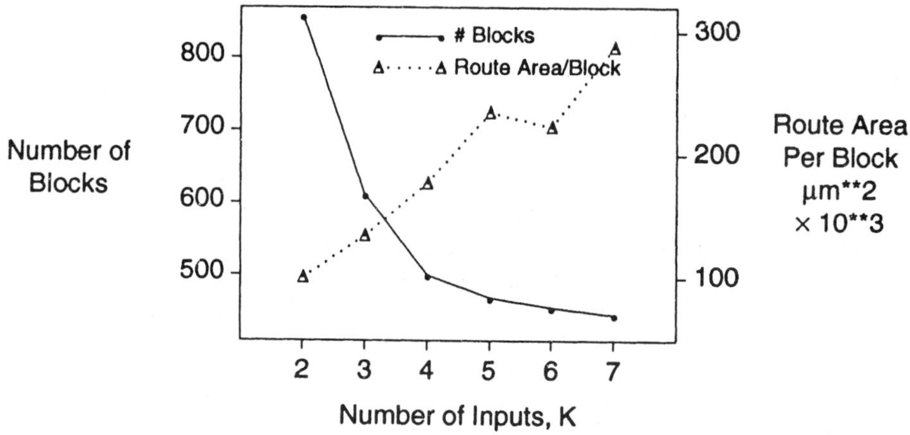

Figure 4.7 - *No. of Blocks and Routing Area / Block, for one Circuit.*

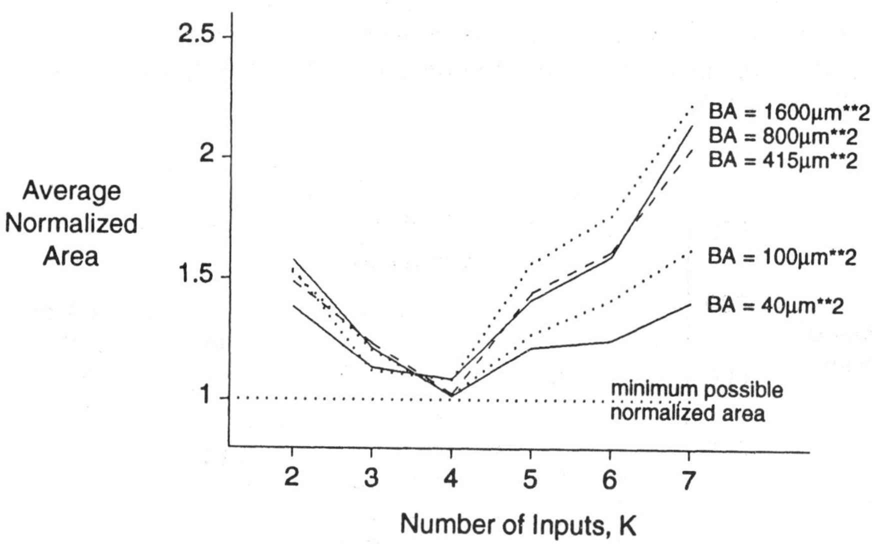

Figure 4.8 - *Average Normalized Total Area for a Single Output K-LUT.*

Figure 4.8 provides two clear conclusions:

1. The most area-efficient value for *K* is approximately 4.
2. This result is largely independent of BA, the reasons for which are discussed in [Rose90c].

4.1.4.2 Multiple-Output, *K*-input Lookup Tables

This section shows similar experimental results for *M*-output *K*-LUTs, where each output is generated by a separate *K*-LUT, and all *K*-LUTs share the same inputs. Figure 4.9 gives a plot of average normalized area versus *K* for LUTs of different size [Koul92a]. Here, the area for each circuit is normalized to the result obtained for K=2 and M=1. This figure confirms the conclusion reached in the previous section, that a 4-LUT is the best choice. Secondly, the plot indicates that it is a poor choice to have more than one output per *K* inputs, since a higher total chip area results. This is probably due to the fact that each output has a high cost, of 2^K memory bits each.

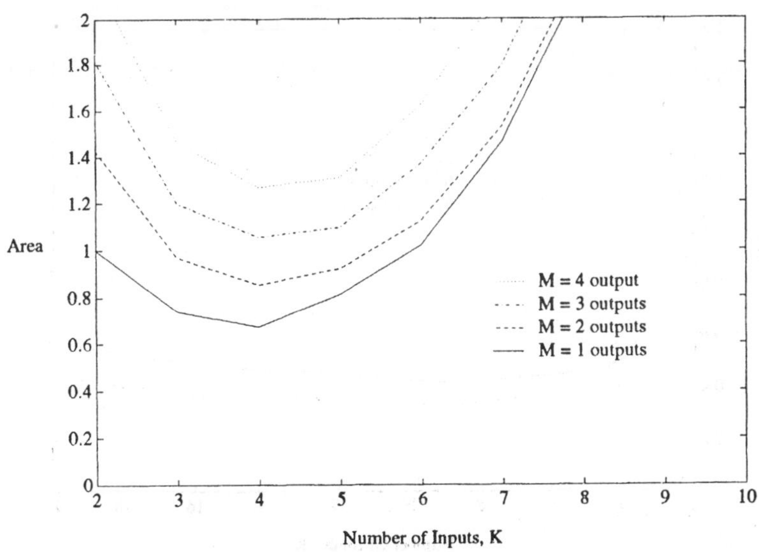

Figure 4.9 - *Average Normalized Total Area [Koul92a].*

4.1.4.3 PLA-based Logic Blocks

Kouloheris and El Gamal observed that the high functionality supplied by a *K*-input lookup table, particularly for *K* larger than five, is not necessary for typical circuits. For this reason, [Koul92a] and [Koul92b] investigated the area-efficiency of PLA-based logic blocks with *K* inputs, *M* outputs, and *N* product terms. The total area required to implement circuits with these blocks was calculated using probability theory. For each *K* and *M*, a good value for *N* was chosen by calculating the mean of a probability density function of the number of product terms for each of a set of benchmark circuits. Using those values of N, Figure 4.10 shows the total area required in the FPGA. Again, each curve is normalized to the result obtained for K=2 and M=1. For this figure, the value of the BA parameter was equivalent to that for an EPROM block [Koul92a]. It can be seen that the PLA that produces the smallest total area has 8 to 10 inputs, 3 or 4 outputs and 12 or 13 product terms. Interestingly, when *K* is greater than about 4, the multi-output PLAs achieve better logic density than the single-output PLAs, in contrast to the lookup tables.

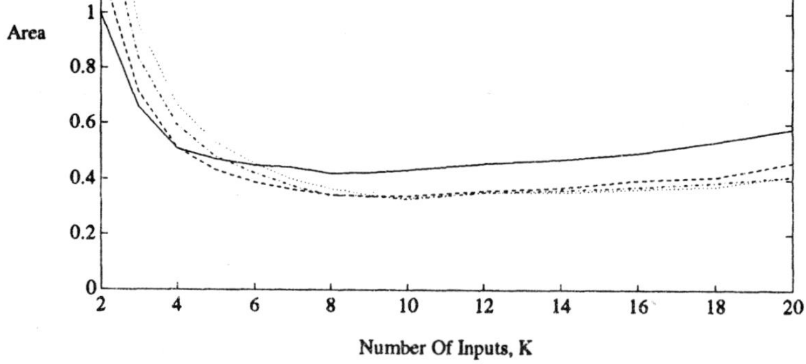

Figure 4.10 - *Average Normalized Total Area [Koul92a].*

Comparing the total area required when using single output 4-LUTs versus PLA-based blocks (with K=10, M=3, and N=12), [Koul92a] shows that the PLA approach requires an average of 4% less area. This may be significant because there are several area-saving optimizations yet to be tried for the PLAs. One possibility is fixing the OR plane (i.e., a PAL-like structure), similar to the architecture found in the Altera FPGAs that were described in Chapter 2.

4.1.4.4 Decomposable Lookup Tables

This section presents one example that illustrates the effects of having decomposable lookup tables, as described in Section 4.1.1. For these results, decomposable blocks with M = 1, 2, 4, or 8 outputs are evaluated. For each of the different logic blocks, the total number of lookup table bits is kept constant at $2^5 = 32$. This means that for the $M = 1$ case, the block has one 5-LUT, for $M = 2$ there are two 4-LUTs (similar to the example that was shown in Figure 4.3), for $M = 4$ there are four 3-LUTs, and for $M = 8$ there are eight 2-LUTs.

A decomposable block requires more area because of the routing area needed to connect to the increased number of pins. On the other hand, a decomposable block is more flexible, so fewer blocks are required to implement a given circuit. This is illustrated in Figure 4.11, which gives the average number of blocks required to implement a set of 15 benchmark circuits, and the area needed per block (including both the logic block area and the routing area), for each of the four values of M. Routing area was calculated using Equation 4.4. The product of the two curves gives the total area needed in the FPGA, which is shown in Figure 4.12. It is apparent that the best area (measured in equivalent bits) is achieved when M is 4. Note that there is a significant gain in moving from a single output block to a two-output block. This decreases the total number of blocks significantly, and more than makes up for an increase in the area required for routing structures due to the increased number of pins.

4.1.4.5 Utility of D Flip-Flops

It is important to determine if having a D flip-flop in the logic block is beneficial. If logic blocks do not have embedded flip-flops then whenever a flip-flop is required by a circuit, it will be necessary to use one or more blocks to implement it. Experimental results from [Rose90c] show that the number of logic blocks needed to implement example circuits increased by a factor of about 2 when the flip-flop was removed from the logic block, depending on the number of flip-flops in the original circuit. However, the logic block size without a D flip-flop is about twice as small, for K in the

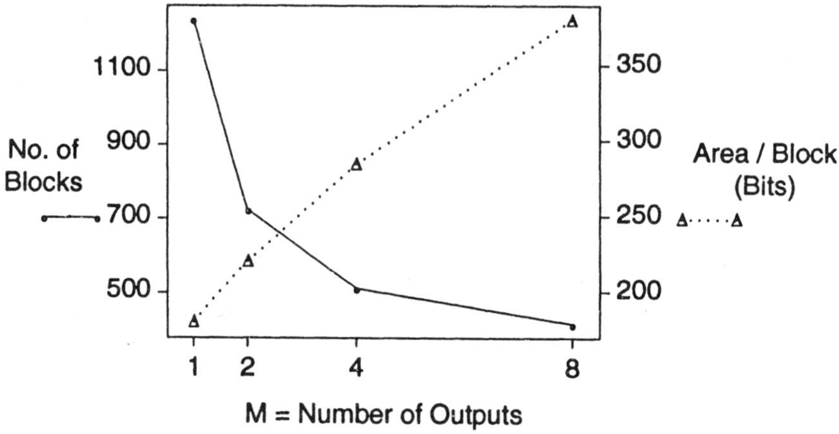

Figure 4.11 - *No. of Decomposable Blocks and Area / Block.*

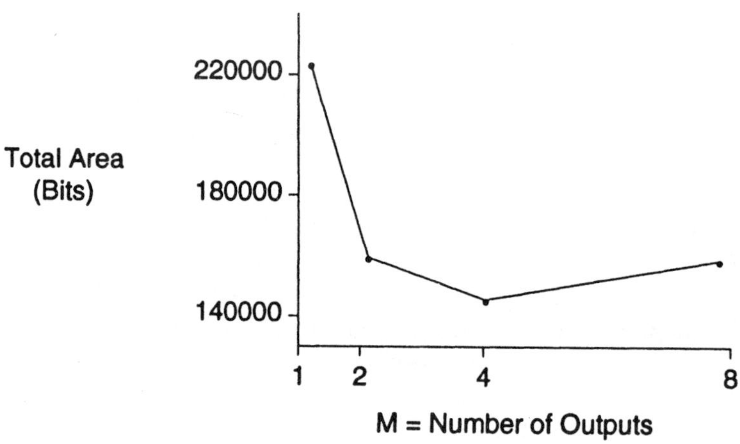

Figure 4.12 - *Total Area for Decomposable Blocks.*

range of 3 to 4. This means that the total logic block area without using a D flip-flop is roughly the same, but because there are about twice as many blocks, the area needed for routing resources will at least double, to realize a given circuit. Since routing area is the dominant part of the overall area, it is always better to include a D flip-flop.

4.1.4.6 Summary of Area Results

The discussion in the previous sections leads to the following conclusions:

1. A 4-input lookup table is the best choice among single-output, K-input lookup tables.
2. Since every connected pin on a logic block incurs a significant penalty for routing area, logic blocks that have a high functionality per pin are area-efficient.
3. Among K-input, multi-output lookup tables for which each output requires an additional 2^K bits, the single-output lookup table is the most area-efficient.
4. For PLA-based logic blocks, the best area is achieved when the PLA has 8 to 10 inputs, 3 or 4 outputs, and 12 or 13 product terms. A multi-output PLA is superior to a single-output PLA, because the expense of additional outputs is small. These blocks appear to be slightly better than the 4-input lookup table.
5. It is beneficial to have multiple outputs in a lookup table-based block if it is decomposable. In particular, rather than a single 5-LUT table, it is better to have the option of two 4-LUTs, and better still to allow four 3-LUTs. However, the increased routing requirements of having eight 2-LUTs is greater than the saving achieved with the greater flexibility and so this final option is inferior.
6. It is advantageous to include a flip-flop in the logic block, because most circuits need sequential logic, and it is expensive to create the flip-flops by using purely combinational logic blocks.

4.2 Impact of Logic Block Functionality on FPGA Performance

The functionality of the logic block has a significant effect on the performance of an FPGA. As functionality increases, the number of levels of blocks required to implement a circuit will decrease because more logic can be implemented in a single block. The delay of each logic block will likely increase, but since there will also be fewer stages of routing, and routing delays in FPGAs are large, the overall delay will likely decrease. This can be illustrated by the following example. Figure 4.13a gives the implementation of the logic function $f = ab\bar{d} + a\bar{b}c + ac\bar{d}$ using two-input NAND gates as the logic blocks. It requires four levels of blocks in the critical path. Figure 4.13b shows an implementation of the same function using 3-input lookup tables, which requires only two levels. The latter involves two fewer stages of routing, and since the programmable interconnect in FPGAs is normally slow, this will likely lead to a significant decrease in delay. However, increasing the functionality of the logic block is likely to increase its combinational delay. The 2-input NAND gate has a delay of about 0.7ns and the 3-input LUT has a delay of 1.4ns, in a 1.2μm CMOS process. Clearly, for a non-zero routing delay between the blocks, the higher functionality of the 3-LUT will result in a faster circuit.

In this section, we will describe an empirical approach and model that have been used to study the effect of the logic block functionality on the performance of an FPGA. It will focus on the research reported in [Sing91a][Sing91b][Sing92] and [Koul91].

4.2.1 Logic Block Selection

As discussed in Chapter 2, there are many kinds of FPGA logic blocks. In [Sing92], many different types of logic blocks were studied by using a simple experimental procedure and abstract model for delays. By contrast, [Koul91] examined a single class of logic blocks (*K*-input lookup tables) using a more complete procedure and a detailed model for delays.

Four classes of logic blocks were selected for comparison in [Sing92]: NAND gates, multiplexers, *K*-input lookup tables and wide AND-OR gate-based blocks. Table 4.1 gives the names of the logic blocks used and describes each one. Two types of NAND gates were considered: a simple

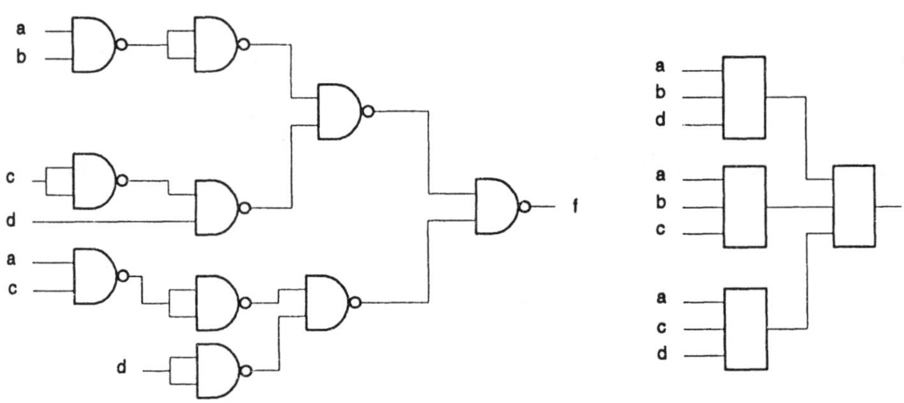

a) Logic Block = 2–input NAND Gate   b) Logic Block = 3–input LUT

Figure 4.13 - *Two Implementations of* $f = ab\bar{d} + abc + ac\bar{d}$.

Block Name	**Logic Function**	**Delay (ns) 1.2μm CMOS**
NAND Gates		
nand2	2-input NAND gate	0.70
nand3	3-input NAND gate	0.88
nand4	4-input NAND gate	1.08
nand2pi	2-input NAND gate with prog inv	1.26
nand3pi	3-input NAND gate with prog inv	1.42
nand4pi	4-input NAND gate with prog inv	1.80
Multiplexers		
mux21	2 to 1 mux	1.08
mux41	4 to 1 mux	1.31
Actel	Actel Act-1 block	1.31
Lookup Tables		
K2	2-input 1-output lookup table	1.39
K3	3-input 1-output lookup table	1.44
K4	4-input 1-output lookup table	1.71
K5	5-input 1-output lookup table	2.03
K6	6-input 1-output lookup table	2.38
K7	7-input 1-output lookup table	2.85
K8	8-input 1-output lookup table	3.26
K9	9-input 1-output lookup table	3.78
AND-OR Gates		
a2o3pi	OR of 3, 2-input product terms	1.88
a4o3pi	OR of 3, 4-input product terms	2.17
a8o3pi	OR of 3, 8-input product terms	2.69
a16o3pi	OR of 3, 16-input product terms	3.77
a32o3pi	OR of 3, 32-input product terms	5.98
a2o5pi	OR of 5, 2-input product terms	1.98
a4o5pi	OR of 5, 4-input product terms	2.27
a8o5pi	OR of 5, 8-input product terms	2.80
a16o5pi	OR of 5, 16-input product terms	3.95
a32o5pi	OR of 5, 32-input product terms	6.05

Table 4.1 - *Logic Block Selection and Delay per Block.*

NAND gate and one that has a programmable inversion capability, in which inputs to the gate can be true or complemented. In the multiplexer class, 2-to-1 and 4-to-1 multiplexers were investigated as well as the Actel ACT-1 logic block. In the lookup table class, *K*-LUTs with a single output were selected, with *K* varying from 2 to 9. Lookup tables were studied in both [Sing92] and [Koul91].

The AND-OR-based blocks that were examined have a structure similar to that in Altera FPGAs. Each of these blocks are described in Table 4.1, using the notation *aKoNpi*, where *K* is the total number of inputs that can be selected to form *N* separate product terms. The product terms are ORed together to generate the output. For example, a8o3pi has eight inputs, each of which can be selected to form three separate product terms that are ORed together. These gates have the programmable inversion capability.

Table 4.1 also gives the worst-case delay for each logic block, determined using the Spice 2G6 circuit simulator [Vlad81], assuming a 1.2μm CMOS process.

4.2.2 Logic Synthesis Procedure

Similar to Section 4.1.2, an experimental procedure is required to take benchmark circuits and "implement" them as FPGAs that use the desired logic blocks. The difference is that the synthesis procedure is directed toward optimizing the performance, rather than the area. Thus, each benchmark circuit is converted into a network of logic blocks while minimizing the number of blocks along the paths between the primary inputs and the outputs of the circuit.

The experimental procedure used in [Sing92] involves first performing technology-independent logic optimization, which was discussed briefly in Chapter 3, and then technology mapping. For the technology mapping step, the best available algorithm for each type of logic block was used. The NAND gates and multiplexers were mapped using the mis 2.2 technology mapper [Detj87]. The lookup tables were mapped in [Sing92] using Chortle-d [Fran91b] which was described in Chapter 3, and in [Koul91] by Chortle-crf [Fran91a] which was also described in Chapter 3. The AND-OR gates were mapped using a mathematical approximation based on the number of inputs to each product, and the number of products. For the study in [Sing92], the synthesis procedure ends at this point. The experimental procedure used in [Koul91] differs from the one described above. In [Koul91], technology mapping is performed first and then placement and routing. The next section describes the way in which the outputs of the experimental procedures are used to measure delays.

4.2.3 Model for Measuring Delay

In [Sing92], the speed of a circuit implemented in an FPGA with a given logic block is a function of the combinational delay of the logic block, D_{LB}, the number of logic blocks in the critical path, N_L, and the delay incurred in the routing between logic blocks, D_R. Assuming that each stage of block incurs one logic block delay and one routing delay, then the total delay, D_{TOT} is

$$D_{TOT} = N_L \times (D_{LB} + D_R) .$$

The value of N_L can be determined for each circuit after the technology mapping step. Each value of D_{LB} is given in Table 4.1.

The value of D_R is more difficult to determine. It is a function of the routing architecture, the fanout of a connection, the length of the connection (which would be determined by the physical placement), the process technology, and the programming technology. In [Sing92] none of these parameters were fixed. Instead, the experimental results below are given as a function of D_R, rather than choosing a specific value for D_R.

[Koul91], in contrast, assumed a specific routing architecture: the row-based architecture present in Actel FPGAs, as described in Chapter 2. The routing switches are characterized as having a switch time constant $T_{SW} = R_{SW} \times C_{SW}$, where R_{SW} is the on-resistance of the switch and C_{SW} is its capacitance. Since full placement and routing was performed, RC networks could be obtained for the interconnections. The analytic expressions in the Rubenstein-Penfield model [Rube83] were then used to calculate approximations for the delays.

4.2.4 Experimental Results

A total of sixteen benchmark circuits were used for the study in [Sing92]. The circuits range in size from 28 to over 700 two-input NAND gate equivalents. Each circuit was passed through the synthesis procedure described in Section 4.2.2, once for every logic block listed in Table 4.1. [Koul91] is also based on many benchmark circuits, of various sizes.

We first present the results for K-input lookup tables since these match well with the area experiments presented in Section 4.1, and because varying K from 2 to 9 represents a broad range of functionality. The subsequent sections cover NAND gates, multiplexers, and wide AND-OR gates respectively, with the last section making a comparison of the best blocks in each of the four classes.

4.2.4.1 Lookup Tables

Figure 4.14 shows the average number of logic block levels in the critical path and the block delay for lookup tables that have from 2 to 9 inputs. The data is averaged over the set of benchmark circuits used in [Sing92]. The figure shows that as functionality (K) increases, the number of logic blocks in the critical path decreases, while the combinational delay of the logic block increases.

The product of the two curves in Figure 4.14 gives the total delay due to the logic blocks. Adding this to the product of the solid curve in the figure and D_R (i.e., the total routing delay) yields the total delay, D_{TOT}. The results of performing these calculations are shown in Figure 4.15, for four values of D_R. Typical values for D_R in ıeal FPGAs range from about 2.5ns to 10ns in 1.2μm CMOS [Vuil91].

For very fast routing delay ($D_R = 0$), the total delay is strictly a function of the number of logic block levels, N_L and the delay of the logic block, D_{LB}. As shown in Figure 4.15, for values of K greater than 2, the total delay, D_{TOT}, is almost constant. This implies that as K is increased above 3, a reduction in delay due to a lower N_L is offset by a higher D_{LB}. Thus for zero routing delay, a 3- or 4-input lookup table is the best, because these have the lowest area.

As D_R increases, the cost in delay of each logic block level begins to dominate, and so the blocks with lower values of N_L achieve superior

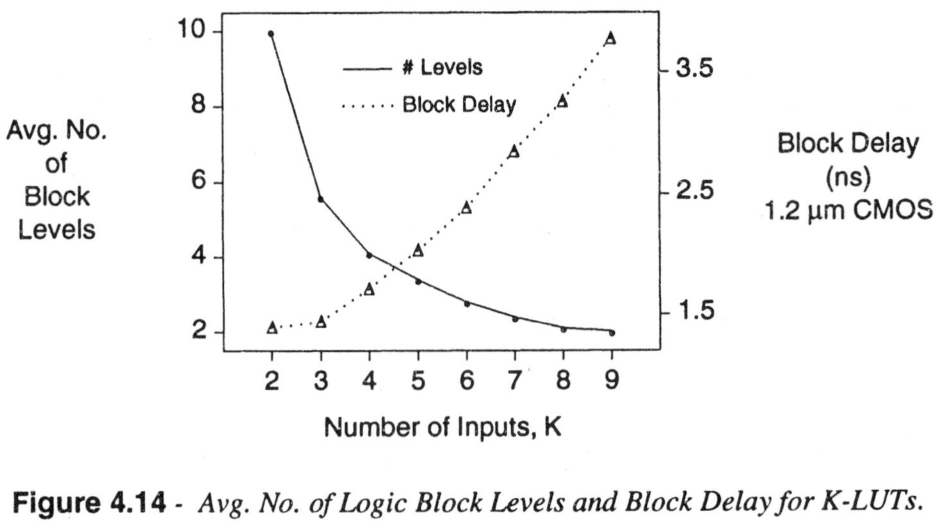

Figure 4.14 - *Avg. No. of Logic Block Levels and Block Delay for K-LUTs.*

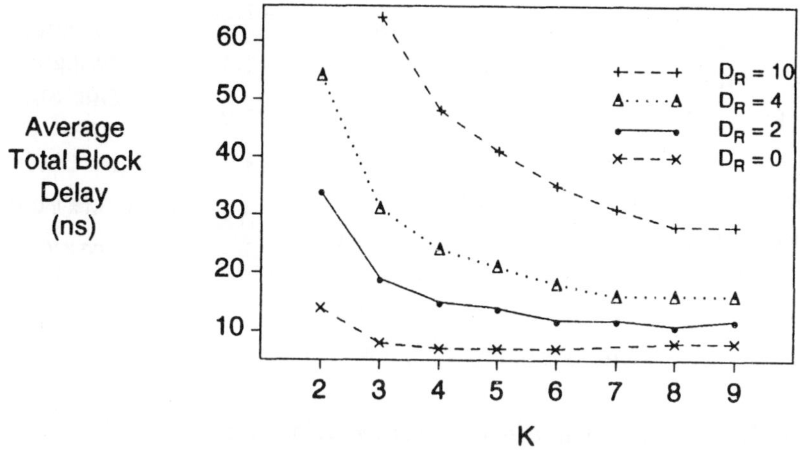

Figure 4.15 - *Average Total Delay for K-LUTs.*

performance. For $D_R=2$ the lowest delay is reached at about the 5- or 6-input cases. For $D_R=4$, the best performance is achieved at about 6 or 7 inputs. The actual choice of a block might be more strongly influenced by the fact that each added input doubles the number of bits in the lookup table, and hence the area. Thus, the 5- and 6-input lookup tables are good choices for $D_R=2$ and $D_R=4$ ns, which are realistic values for routing delay.

As D_R increases to 10 ns, the best value of K continues to increase, to about $K=8$. It is clear that for large routing delays it is advantageous to have highly functional logic blocks, because the number of logic block levels will be reduced. The advantage stems from the fact that the routing delay far exceeds the logic block delay in this case.

The results reported in [Koul91] are very similar to those shown in Figure 4.15, even though the synthesis procedure continued through placement and routing, and a full RC-network delay was calculated. This serves to strengthen the results from [Sing92], since they were based on more abstract models.

4.2.4.2 NAND Gates

Figure 4.16 shows the results for the NAND gates. It gives the average number of logic levels in the critical path (the solid lines) and the delay of each block (the dotted lines) for 2-, 3- and 4-input NAND gates. The curves marked with triangles are for gates with programmable inversion and the

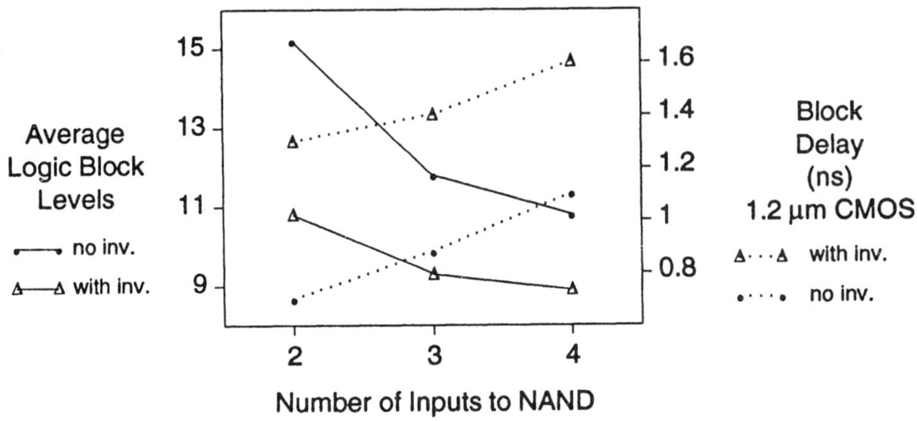

Figure 4.16 - *Avg. Logic Block Levels and Block Delay for NANDs.*

curves marked with bullets do not have programmable inversion. It is clear from the figure that the programmable inversion feature significantly reduces the number of blocks in the critical path. However, the programmable inversion increases the delay per logic block, by about 0.6 ns.

Figure 4.17 gives the total delay for the NAND gates. It shows that, for all but $D_R = 0$, the NAND gates with programmable inversion give better performance than the NAND gates without this feature. This is because at

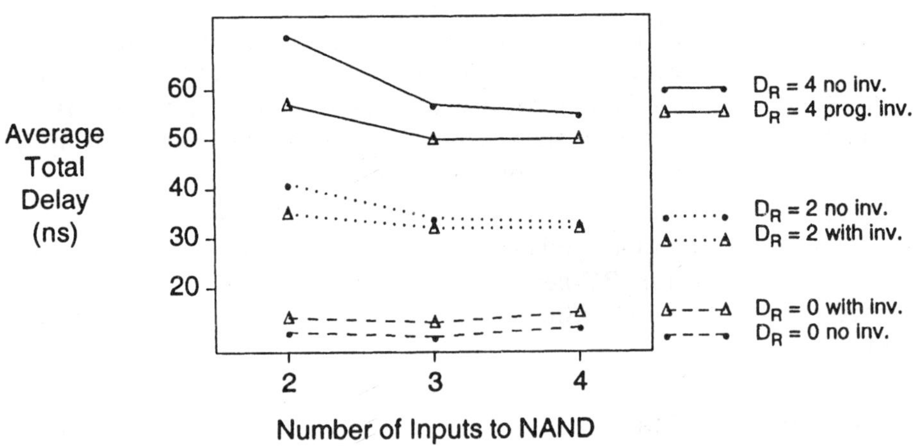

Figure 4.17 - *Average Total Delay for NAND Blocks.*

higher routing delays the difference in gate delays is more than compensated for by the saving in the number of levels. Only for $D_R=0$ do the NAND gates without programmable inversion yield better performance.

The figure also suggests that there is little or no improvement beyond a 3-input NAND gate, which means that the reduction in the number of levels does not compensate for the increased block delay.

4.2.4.3 Multiplexer-Based Blocks

The experimental data for the three multiplexer configurations is given in Table 4.2. The first column names the gate, the second lists the combinational delay from Table 4.1, the third gives the average number of logic blocks in the critical path, N_L, over all 16 circuits, and the fourth column contains the standard deviation of this average. Columns 5 through 8 give the total delay, D_{TOT}, for different values of the routing delay, D_R. The Actel logic block exhibits the lowest N_L. This is due to the high number of logic functions that this block can perform. The combinational delay of the Actel logic block is the same as that of a 4-to-1 multiplexer (mux41), and because it leads to a lower N_L, it gives better performance for all values of D_R.

Logic Block	D_{LB} **(ns)**	N_L	**St. Dev.**	$D_{TOT} = N_L \times (D_{LB} + D_R)$			
				$D_R=0$ (ns)	$D_R=2$ (ns)	$D_R=4$ (ns)	$D_R=10$ (ns)
mux21	1.1	9.9	4.7	11	30	50	110
mux41	1.3	6.1	2.3	8	20	33	69
Actel	1.3	4.4	2.0	6	15	23	50

Table 4.2 - *Avg. Critical Path Length & Total Delay for Multiplexers.*

4.2.4.4 AND-OR Gates

Figure 4.18 presents the results for the wide AND-OR gates. For these results, the values 3 and 5 have been chosen for the number of product terms, N. The figure shows that the 5-product term block provides a significant decrease in the number of logic levels over the 3-product term case, yet its combinational delay is only slightly larger, for all values of K from 2 to 32. Since the 5-product term blocks are superior to the 3-product term blocks, we will give the total delay results only for $N=5$.

The total delay for the aKo5pi block is plotted in Figure 4.19. As before, for low values of D_R the lower functionality blocks are superior, and the higher functionality blocks perform well for higher values of D_R. When

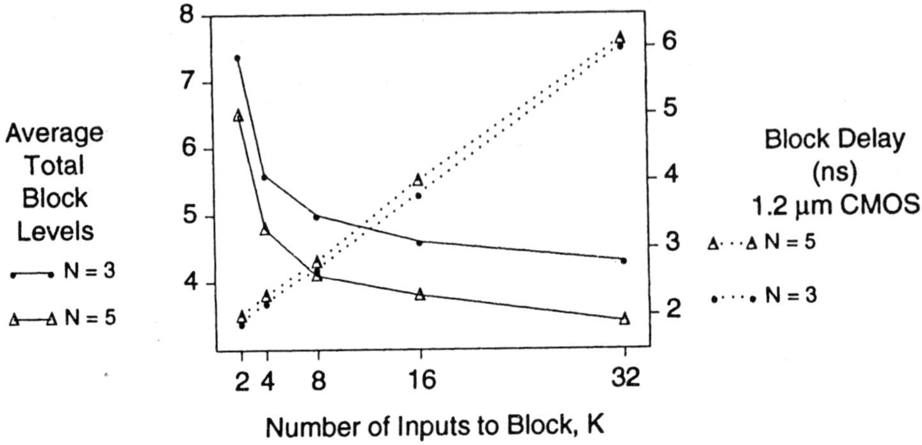

Figure 4.18 - *Avg. Block Levels and Block Delay for Wide AND-ORs.*

longer routing delays are assumed, then the blocks with greater K become more attractive.

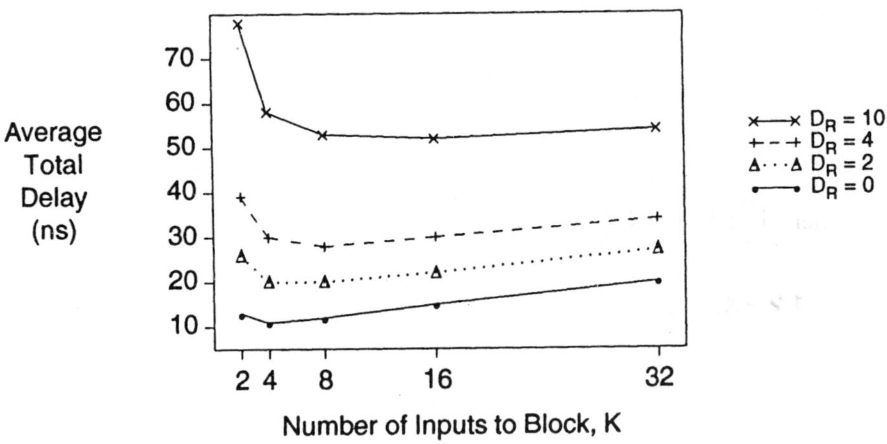

Figure 4.19 - *Average Total Delay aKo5pi Wide AND-ORs.*

4.2.4.5 Overall Comparison

Figure 4.20 gives the total delay of the best logic blocks from each class as a function of D_R. More details for each block are provided in Table 4.3. The table gives the individual values of D_{LB} and N_L for each block, the standard deviation of N_L, and the D_{TOT} for each value of D_R. An interesting conclusion from this data is that the fine-grain logic blocks, such as the 2-

input and 3-input NAND gates (even with programmable inversion) exhibit markedly lower performance than any other class of logic blocks. This is a significant conclusion, given that some commercial FPGAs use the two-input NAND gate as the basic logic block. Note that the result is true even for a routing delay of zero, which provides an interesting perspective on mask-programmed architectures. They currently use NAND gates as their basic block, but should perhaps use a higher functionality block, as suggested in [ElGa89a].

At zero routing delay, the Actel logic block is the fastest because it has a very small combinational delay, combined with a low number of logic block levels.

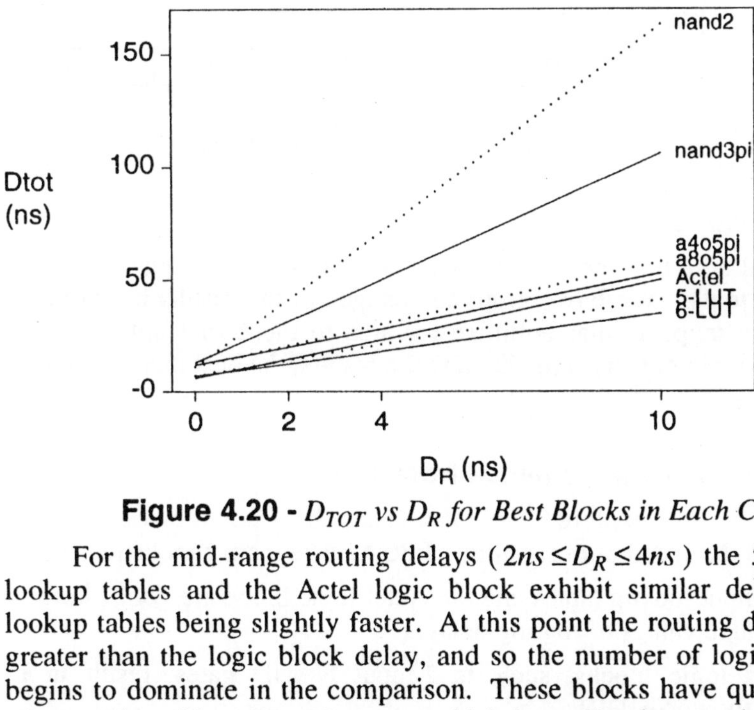

Figure 4.20 - *D_{TOT} vs D_R for Best Blocks in Each Class.*

For the mid-range routing delays ($2ns \le D_R \le 4ns$) the 5- and 6-input lookup tables and the Actel logic block exhibit similar delays, with the lookup tables being slightly faster. At this point the routing delay is mostly greater than the logic block delay, and so the number of logic block levels begins to dominate in the comparison. These blocks have quite low values of N_L. The wide AND-OR gates, which have N_L close to the Actel block, exhibit worse performance because of a significantly higher combinational delay.

For large delays ($D_R = 10ns$) the 5- and 6-input lookup tables are significantly faster. This is because here the only important factor is the number of logic levels, and as Table 4.3 shows, the lookup tables have significantly lower values of N_L. Notice that the wide AND-OR gates do not approach this level. It is possible, however, that improved technology mapping tools could enhance the results for these blocks, as discussed in [Sing92].

Logic Block	D_{LB} (ns)	N_L	St. Dv.	$D_{TOT} = N_L \times (D_{LB} + D_R)$			
				$D_R = 0$ (ns)	$D_R = 2$ (ns)	$D_R = 4$ (ns)	$D_R = 10$ (ns)
nand2	0.70	15.2	5.8	11	41	71	163
nand3pi	1.4	9.3	3.8	13	32	50	106
a4o5pi	2.3	4.8	1.2	11	20	30	58
a8o5pi	2.8	4.1	1.2	12	20	28	53
Actel	1.3	4.4	2.0	6	15	23	50
K5	2.0	3.4	1.2	7	14	21	41
K6	2.4	2.8	1.0	7	12	18	35

Table 4.3 - *Overall Comparison of Critical Path Length and Total Delay.*

4.2.4.6 Limitations of Results

We should note that these results depend heavily on the quality of the logic synthesis tools. We have observed changes in the results by moving from technology mappers that optimize for area to those that optimize for delay. In the experiments used to generate the results, the best mapping tools available have been used.

4.2.4.7 Summary of Performance Results

This section has explored the relationship between the logic block architecture and the speed of the resulting FPGA. The main conclusions are:

1. 5- and 6-input lookup tables and the Actel logic block are good choices for mid-range values of routing delay.
2. Fine-grain logic blocks, such as 2-input NAND gates, result in a significantly worse delay.
3. The programmable inversion capability on the inputs of small gates, such as NAND gates, improves their performance.
4. For wide AND-OR gates, blocks with five product terms exhibit superior performance over three product terms.

It seems that the wide AND-OR gates do not achieve performance comparable to the best blocks, but it is possible that better logic synthesis for these blocks would lead to improved performance.

4.3 Final Remarks and Future Issues

A clear conclusion from the research discussed in this chapter is that fine-grain logic blocks are a poor choice in terms of both area and performance. The reason is that they require too many stages of routing, and routing structures in FPGAs are both large and slow. On the other hand, too much functionality can be a disadvantage as well, since the results of Section 4.1.4.1 show that having an excessive number of pins connected to a block results in greater total chip area. Thus, a high functionality per pin is a definite advantage, which is why lookup tables appear to be a good choice for a logic block. It was shown that the best value for K in a K-input lookup table is 4.

There are many possibilities in logic block architecture that are worth exploring. Some of these are:

1. Further investigation of PLA-based structures may show that these blocks have a significant advantage in terms of total area.
2. Decomposable lookup tables, discussed in section 4.1.4.4, warrant further investigation.
3. Non-homogeneous arrays of logic blocks may offer better performance versus area tradeoffs than do homogeneous arrays.
4. Hierarchical organizations of FPGAs may be better than flat FPGAs.

The material presented in this chapter is based on early research attempts to study experimentally the effects of logic block architecture on the logic density and performance of FPGAs. The reported results are only as good as the CAD tools used to generate them. The development of superior tools will undoubtedly lead to better assessment of the architectural choices and probably to better architectures.

CHAPTER 5
Routing for FPGAs

In Chapter 3, technology mapping in CAD systems for FPGAs was discussed in detail. The next step in such systems is the *placement* of the logic blocks. This problem in the FPGA environment is very similar to placement tasks for other technologies, for example standard cells. A number of efficient techniques for placement have already been developed and well documented in the technical literature [Hana72] [Sech87]. Since these techniques can easily be adapted to use for FPGAs, we will not pursue the placement task in this book. This chapter focuses on the next step in the CAD system, where the *routing* of interconnections among the logic blocks is realized. As Figure 5.1 indicates, routing is the final phase of a circuit's implementation, after which the FPGA can be configured by a programming unit.

The routing algorithms that are appropriate for an FPGA depend on its routing architecture. Algorithms for two different architectures are examined in this chapter: one that has only horizontal routing channels (row-based FPGA), and one that has both vertical and horizontal routing channels (symmetrical FPGA). In terms of the FPGAs described in detail in Chapter 2, the first type corresponds to an Actel architecture, while the second resembles a Xilinx chip. Note that the routing issues for PLD-like FPGAs, such as those offered by Altera, are not discussed in this chapter. In these architectures routing is simple, because they contain uniform interconnection structures that provide complete connectivity. Before proceeding to specific routing algorithms, we will first define some common routing terminology and then give an overview of the routing strategy that is used.

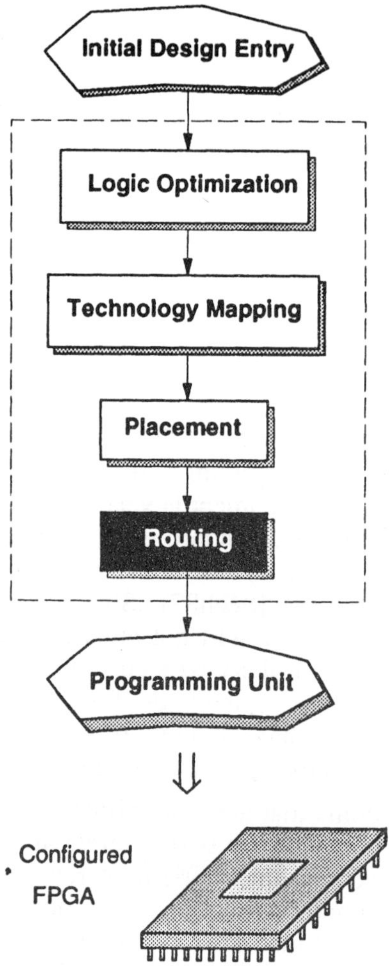

Figure 5.1 - *A Typical CAD System for FPGAs.*

5.1 Routing Terminology

Software that performs automatic routing has existed for many years, the first algorithms having been designed for printed circuit boards. Over the years there have been many publications concerning routing algorithms, so that the problem is well defined and understood. The following list defines the routing terminology that is used throughout this chapter:

- *pin* - a logic block input or output.
- *connection* - a pair of logic block pins that are to be electrically connected.
- *net* - a set of logic block pins that are to be electrically connected. A net can be divided into one or more connections.
- *wire segment* - a straight section of wire that is used to form a part of a connection.
- *routing switch* - a programmable switch that is used to electrically connect two wire segments.
- *track* - a straight section of wire that spans the entire width or length of a routing channel. A track can be composed of a number of wire segments of various lengths.
- *routing channel* - the rectangular area that lies between two rows or two columns of logic blocks. A routing channel contains a number of tracks.

5.2 General Strategy for Routing in FPGAs

Because of the complexity involved, the solution of large routing problems, such as those encountered in FPGAs, usually requires a "divide and conquer" strategy. This approach is extolled in the three-step process described in [Loren89]:

1. Partition the routing resources into routing areas that are appropriate for both the device to be routed and the routing algorithms to be employed.
2. Use a *global router* to assign each net to a subset of the routing areas. The global router does not choose specific wire segments and routing switches for each connection, but rather it creates a new set of restricted routing problems.
3. Use a *detailed router* to select specific wire segments and routing switches for each connection, within the restrictions set by the global router.

The advantage of this approach is that each of the routing tools can more effectively solve a smaller part of the routing problem. More specifically, since the global router need not be concerned with allocating wire segments or routing switches, it can concentrate on more global issues, like balancing the usage of the routing channels. Similarly, with the reduced number of detailed routing alternatives that are available for each connection because of the restrictions introduced by the global router, the detailed router can focus on the problem of achieving connectivity.

This strategy has been adopted for routing in both types of FPGAs that are discussed in this chapter. The global router first selects routing channels for each connection. Then, within the constraints imposed by the global router, the detailed router implements each connection by choosing specific wire segments and routing switches. It will be apparent that the global routing issues are similar in both row-based and symmetrical FPGAs, but the detailed routing problems warrant substantially different algorithms.

5.3 Routing for Row-Based FPGAs

As depicted earlier, in Figure 2.1, a row-based FPGA consists of rows of logic blocks that are separated by horizontal routing channels. This section describes routing algorithms that have been developed specifically for this class of FPGAs [Green90].

An FPGA based on the Actel architecture is illustrated in Figure 5.2. As shown, the routing channels consist of horizontal wire segments of various lengths, separated by routing switches. Adjacent wire segments can be joined together, allowing longer segments to be formed where necessary. Dedicated vertical segments are attached to the logic block pins and can be connected via routing switches to any horizontal wire segments that they cross. There are also other vertical segments that serve as "feed-throughs" from one routing channel to another. Note that Figure 5.2 shows only the dedicated vertical segments for the logic block marked by "*". Also, only three vertical feed-throughs are shown to avoid cluttering of the figure.

The connectivity illustrated in Figure 5.2, specifically the fact that all horizontal and vertical wire segments that cross can be connected, allows the routing problem to be partitioned into individual channels. This resembles classic channel routing in any row-based architecture, except that in classic channel routing wire segments can be placed freely wherever they are needed, whereas the wire segments in an FPGA are fixed in place before routing is performed. It is still possible to use classic routing algorithms for some special cases [Green90], but in general a new approach is required.

As indicated in Section 5.2, the first step in the routing process is global routing. This entails first dividing each multi-pin net into a set of connections, and then assigning each connection to a specific routing channel. The global router can use the vertical feed-through segments for nets that span multiple channels. Since global routing in this context is very similar to that in other row-based technologies, such as standard cells, existing algorithms can be used. Good global routing techniques have been widely published for those technologies [Loren89] [Sech87] [Cong88] [Rose90a], so they will not be described here.

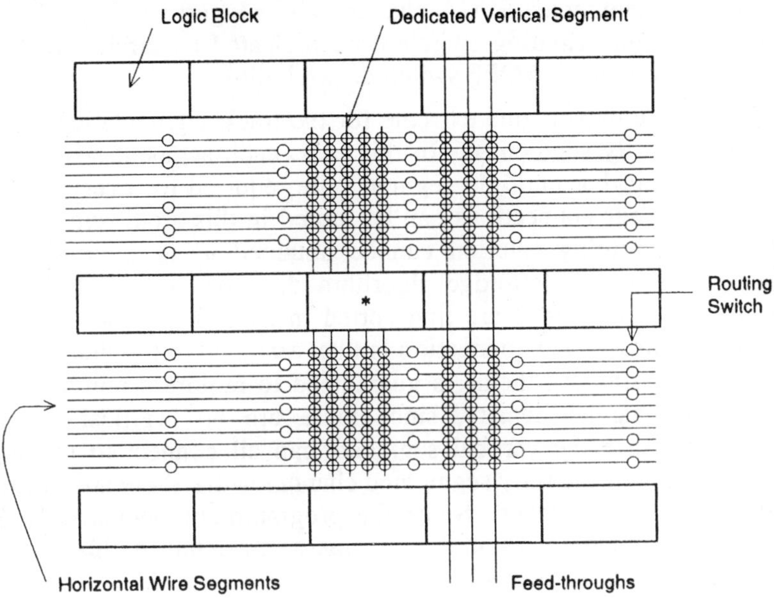

Figure 5.2 - *A Row-Based FPGA.*

5.3.1 Introduction to Segmented Channel Routing

After global routing, each channel can be considered as a separate *detailed* routing problem. A channel will contain some number of connections, each one involving logic block pins or vertical feed-throughs. The task of the detailed routing algorithm is to allocate wire segments for each connection in a way that allows all connections to be completed. In addition, it may be necessary to minimize the routing delays of the connections. This can be accomplished by limiting the total number and length of segments used by a connection. The appropriate algorithm depends on the segmentation of the tracks and the requirements for minimizing delays, as discussed below.

Figure 5.3a shows an example of a routing problem in which the connections called C_1 to C_4 are to be routed. Figure 5.3b indicates how the connections might be routed in a mask-programmed channel, where there is complete freedom for placing the wire segments. The channel is divided into columns, as shown by the vertical segments in the figure. Some columns represent logic block pins and others are vertical feed-throughs. As the column labels in the figure indicate, each of C_1 to C_4 specifies two columns that are to be connected. Figures 5.3c - 5.3f present several different scenarios for a segmented FPGA channel, and suggest routing solutions for each segmentation. Routing switches are indicated by circles, with an ON switch drawn solid and an OFF switch drawn hollow.

Figure 5.3c depicts one extreme for the track segmentations, in which every track is fully segmented. In this case, each segment spans only one column, meaning that multiple segments are required for every connection. As shown, the four connections can be routed in this architecture with only two tracks. A routing solution can be obtained using a straight-forward approach, such as the left-edge algorithm described in [Hash71]. In this algorithm, the connections are first sorted in ascending order according to their leftmost pins. Each connection is then assigned to the first track that is available. This simple scheme is guaranteed to require a number of tracks equal to the channel density, since there are no "vertical constraints" [Loren89] (because the vertical segments can all connect to every track). The shortcoming of the fully segmented channel is that there are an excessive number of routing switches. Since the programmable switches in FPGAs always have significant resistance and parasitic capacitance, this would cause relatively large routing delays.

The opposite extreme to full segmentation is depicted in Figure 5.3d, which shows a channel in which each track contains only one segment for its entire length. In this case, the number of tracks required for routing is always equal to the number of connections. No routing algorithm is necessary since any choice of track for a connection will do. The problem with this segmentation is that excessive area is required for the large number of tracks, and each connection will be subjected to a large capacitive load due to the long segments.

An intermediate approach to channel segmentation is illustrated in Figure 5.3e, where the tracks have segments of various lengths. Each connection must be routed in a single segment, since no switches are available where segments in the same track meet. This 1-segment problem is a special case of segmented routing and can be solved using the simple algorithm described in Section 5.3.2.

Additional flexibility in the channels can be added by allowing segments that abut to be joined by switches, as depicted in Figure 5.3f. This implies that connections can occupy more than one segment and greatly increases the complexity of the problem. A general algorithm for this class of segmented routing problem is presented in Section 5.3.3.

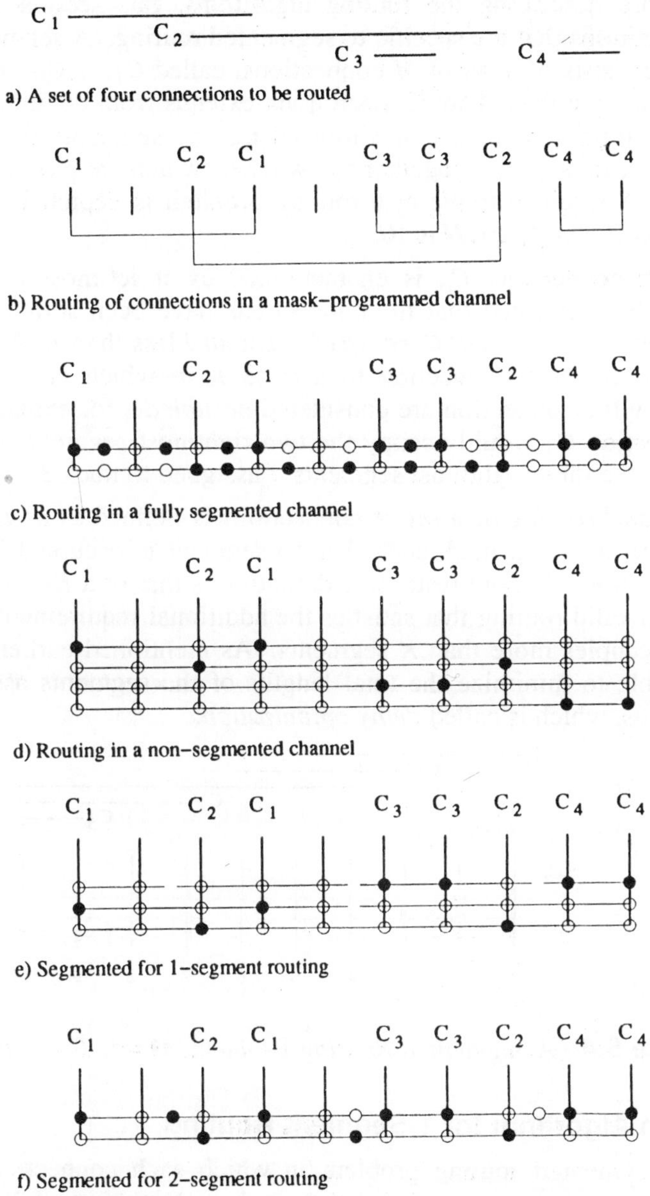

Figure 5.3 - *Examples of Segmented Channels.*

5.3.2 Definitions for Segmented Channel Routing

Before describing the routing algorithms, this section first presents some definitions that are specific to segmented routing. A segmented routing problem consists of a set of M connections, called $C_1, ..., C_M$, and a set of T tracks numbered from 1 to T. Each track extends from column 1 to column N and comprises segments of various lengths. Segments that abut in the same track can be joined together by switches, which are placed between two columns. A small example of a routing problem is depicted in Figure 5.4, where M is 5, T is 3, and N is 10.

Each connection, C_i, is characterized by it leftmost column, called $left(C_i)$. It is assumed that the connections have been sorted according to their left ends, so that $left(C_i) \leq left(C_j)$ for all i less than j. A routing algorithm can *assign* a connection to a track t, in which case the segments spanned by the connection are considered *occupied*. For example, in Figure 5.4 connection C_3 would occupy the two rightmost segments if assigned to track 1, or the three rightmost segments if assigned to track 3.

A *valid routing* of a set of connections is defined as an assignment of each connection to a track such that no segment is occupied by more than one connection. A more restrictive definition is that of a *K-segment routing*, which is a valid routing that satisfies the additional requirement that no connection occupies more than K segments. As mentioned earlier, it may also be desirable to minimize the total lengths of the segments assigned to the connections, which is called *delay optimization*.

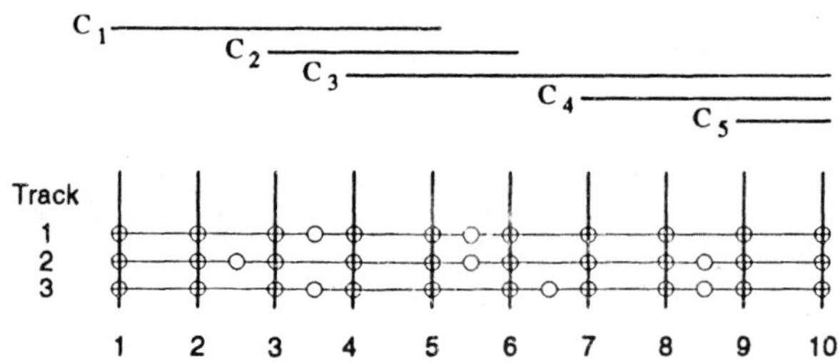

Figure 5.4 - *A Segmented Routing Problem, $M=5$, $T=3$, and $N=10$.*

5.3.3 An Algorithm for 1-Segment Routing

A segmented routing problem in which each connection should use only one segment is a special case that can be solved by the following greedy algorithm. Assume that there are M connections and T tracks. Assign the connections in order of increasing left ends as follows. For each connection, find the set of tracks in which the connection would occupy one segment.

Eliminate any tracks in which this segment is already occupied. From the remaining tracks, assign the connection to the one whose rightmost end is furthest to the left. This simple scheme is guaranteed to find a solution for any set of connections in any segmented channel, if a solution exists. Since it is necessary to check each track for each connection, the run-time of the algorithm would be $O(MT)$. An example of a 1-segment routing problem was illustrated in Figure 5.3e. Note that for the track segmentations and connections that are shown in Figure 5.4, a valid 1-segment routing is not possible.

5.3.4 An Algorithm for K-Segment Routing

For K greater than 1, it has been shown [Green90] that K-segment routing belongs to the class of problems known as NP-complete. This section presents an algorithm that is guaranteed to find a solution for a K-segment problem, if a solution exists. The algorithm works by building a data structure, called an *assignment tree*, to represent the effect of optionally assigning each connection to each track. After the assignment tree is completely constructed, a routing solution can be read directly from it.

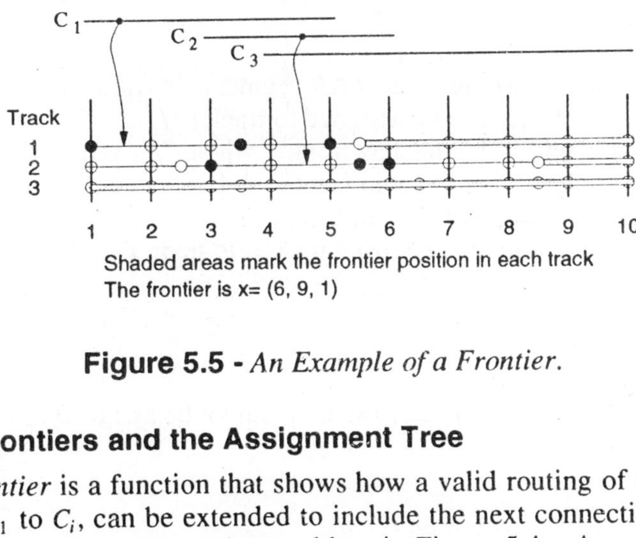

Figure 5.5 - *An Example of a Frontier.*

5.3.4.1 Frontiers and the Assignment Tree

A *frontier* is a function that shows how a valid routing of a set of connections, C_1 to C_i, can be extended to include the next connection, C_{i+1}. As an example, consider the routing problem in Figure 5.4. Assume that connections C_1 and C_2 have been assigned to tracks 1 and 2, respectively, as depicted in Figure 5.5. The frontier marks the leftmost column at which each track is still unoccupied, as indicated by the shaded areas in the figure. It is apparent that connection C_{i+1} can be assigned to a track if the frontier has not advanced past $left(C_{i+1})$. In Figure 5.5, C_3 can be assigned only to track 3.

Given a valid routing of $C_1, ..., C_i$, the frontier, x, can be specified by the T-tuple $(x[1], x[2], ..., x[T])$, where $x[t]$ is the leftmost column in track t in which the segment present in that column is not occupied. The T-tuple then provides enough information to determine which tracks are available for connection C_{i+1}. For the special case of $i=0$, define the initial frontier $x_0=(0,0,...,0)$. For $i=M$, define the final frontier x_M.

Frontiers are used to build a data structure called an *assignment tree*, which is a graph that keeps track of partial routing solutions. Level 0 of the assignment tree represents the frontier x_0. A node at level i corresponds to a frontier resulting from some valid routing of $C_1, ..., C_i$. The assignment tree has a maximum of M levels, one for each connection. If a valid routing of all M connections exists, then level M of the assignment tree will contain a single node corresponding to x_M. Otherwise, level M will be empty.

The assignment tree is the heart of the K-segment routing algorithm. Given level i of the tree, level $i+1$ can be constructed inductively as follows [Green90]:

```
for each node x_i in level i {
   for each track t, 1 ≤ t ≤ T {
      if x_i[t] ≤ left(C_{i+1}) {
         if C_{i+1} would occupy ≤ K segments in track t {
            /* C_{i+1} can be assigned to track t */
            let x_{i+1} be the new frontier when C_{i+1} is assigned to track t
            if x_{i+1} is not yet in level i+1 {
               add node x_{i+1} to level i+1
               add an edge from node x_i to node x_{i+1}. Label it with t
            }
         }
      }
      else {
         /* x_i[t] > left(C_{i+1}) so C_{i+1} cannot be assigned to track t */
      }
   }
}
```

The above procedure is applied once for each of the M connections to be routed. Once level M has been reached, a valid routing can be obtained by retracing the path from node x_M to node x_0. If no nodes are added to the assignment tree for some level $i+1$ then no valid routing of the connections exists.

As mentioned earlier, it may also be desirable for the routing algorithm to optimize the total lengths of the segments used by the connections. With a minor variation, the assignment tree can solve this problem as well. Each edge, labeled t, should be assigned an additional label, w, which represents the *weight* of the edge and corresponds to the total length, measured in columns spanned, of all segments assigned to $C_1, ..., C_{i+1}$. The construction of the assignment tree is modified as follows. When considering track t for C_{i+1}, if a search at level $i+1$ finds that the new node x_{i+1} already exists, examine the weight of its current incoming edge relative to weight w. If the latter is smaller, replace the edge entering x_{i+1} with edge t from x_i. Using this scheme, the solution traced back from node x_M will correspond to a minimal weight routing.

5.3.4.2 An Example of an Assignment Tree

An example of an assignment tree for the routing problem in Figure 5.4 is given in Figure 5.6. The figure shows a valid 2-segment routing, with delay optimization. The frontiers are given inside the nodes of the tree. Each edge has a two-part label (t, w). This corresponds to an assignment of C_{i+1} to track t, with a cumulative length of all segments used for $C_1, ..., C_{i+1}$ of w. The routing solution that can be traced back from node x_M is shown by the bold edges. In the figure, an edge marked with an S has been abandoned because it represents the assignment of a connection to a track in which the connection would occupy more than two segments. An edge marked with * is abandoned because it leads to a frontier that is already present in level $i+1$.

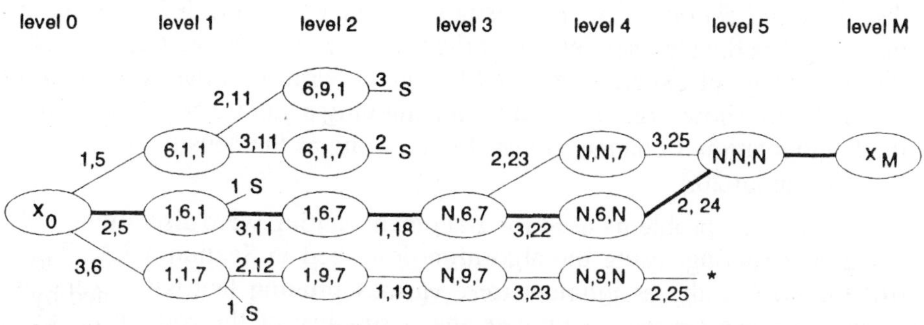

Figure 5.6 - *An Example of an Assignment Tree.*

5.3.4.3 Run-time of the K-Segment Routing Algorithm

The run-time of the K-segment routing algorithm can be determined by noting that for a connection C_{i+1}, each track must be considered for each node at level i. This involves checking each of the T-tuple frontiers in each node, leading to a run-time of $O(MLT^2)$, where L is the maximum number of nodes per level. For a K-segment routing, it can be shown [Green90] that $L \leq (K+1)^T$, which yields a run-time of $O(MT^2(K+1)^T)$. Thus, even though K-segment routing is NP-complete, the algorithm shown above has a linear run-time in the number of connections, M, when the number of tracks, T, is fixed.

5.3.5 Results for Segmented Channel Routing

This section presents some results of experiments with segmented channel routing [Green90]. Two different segmented channels are used. In one channel, called Segmentation-1, the lengths of the segments are tuned so as to give a high probability of achieving a 1-segment routing of a set of connections. In the other channel, called Segmentation-2, the segments are tuned for 2-segment routing. The connections that are to be routed are chosen from a probability distribution that specifies both the starting point of a connection and its length. This distribution is based on real placements from 510 channels in 34 FPGA designs. In all cases, the routing channel has 32 tracks and 40 columns.

The experiments were conducted as follows. First, a set of connections was selected from the probability distribution. A variable number of connections was used for each experiment so that a range of channel densities would be investigated. The Segmentation-1 problems were routed using the scheme described in Section 5.3.3. The results are given in Figure 5.7. The horizontal axis gives the channel density of the problems, and the vertical axis shows the percentage of experiments tried for which valid routings were achieved. As the figure shows, the probability of achieving a valid 1-segment routing in the Segmentation-1 channel is very high as long as the density of the channel is below about 20.

The same problems used for 1-segment routing were also attempted for 2-segment routing, using the algorithm described in Section 5.3.4. To control the size of the assignment tree, various pruning heuristics were used. Although these heuristic eliminate the optimality of the routing algorithm, the results obtained are still excellent. As Figure 5.7 shows, valid 2-segment routings in Segmentation-2 can be obtained with a high probability for densities approaching 29.

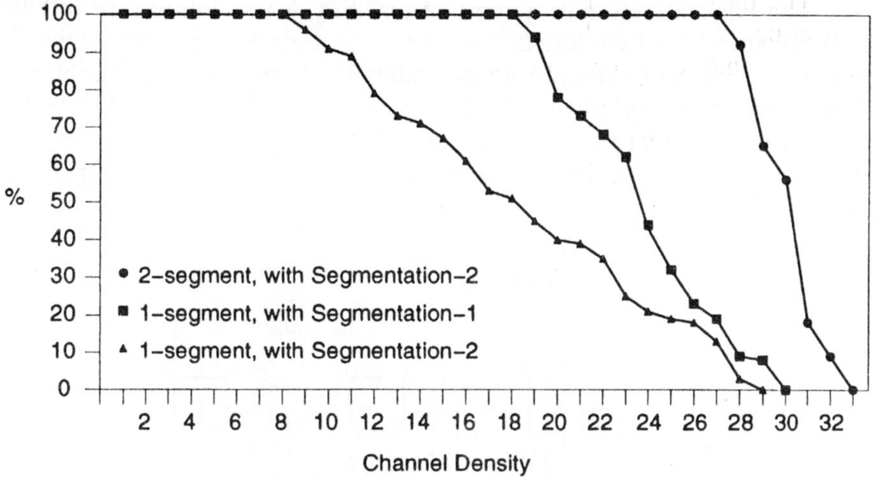

Figure 5.7 - *Segmented Routing Results.*

Figure 5.7 also provides the results of an additional experiment, that of attempting 1-segment routing in Segmentation-2. As shown, for densities above about 8, 1-segment routings are difficult to achieve using Segmentation-2. This shows that it is important to tune the segmentation of the tracks for the type of K-segment routing that is desired.

5.3.6 Final Remarks for Row-Based FPGAs

This section has presented some of the important issues for routing in row-based FPGAs, such as those offered by Actel Corp.. It has been shown that the routing problem in these chips can be divided into independent channels. Each channel then represents a separate segmented channel routing problem. Algorithms have been described that can achieve valid routings in these channels, even when the density is quite close to the total number of tracks.

The next section focuses on a different FPGA routing problem, that for symmetrical FPGAs like those offered by Xilinx. It will be apparent that these architectures present significantly different problems than those in row-based FPGAs.

5.4 Routing for Symmetrical FPGAs

The basic structure of a symmetrical FPGA was introduced in Chapter 2. It consists of a two-dimensional array of logic blocks interconnected by both horizontal and vertical routing channels. Like row-based architectures, global routing for symmetrical FPGAs is similar to that in other technologies, such as standard cells. For this reason, most of the discussion in this section will center on detailed routing, with global routing being discussed only briefly.

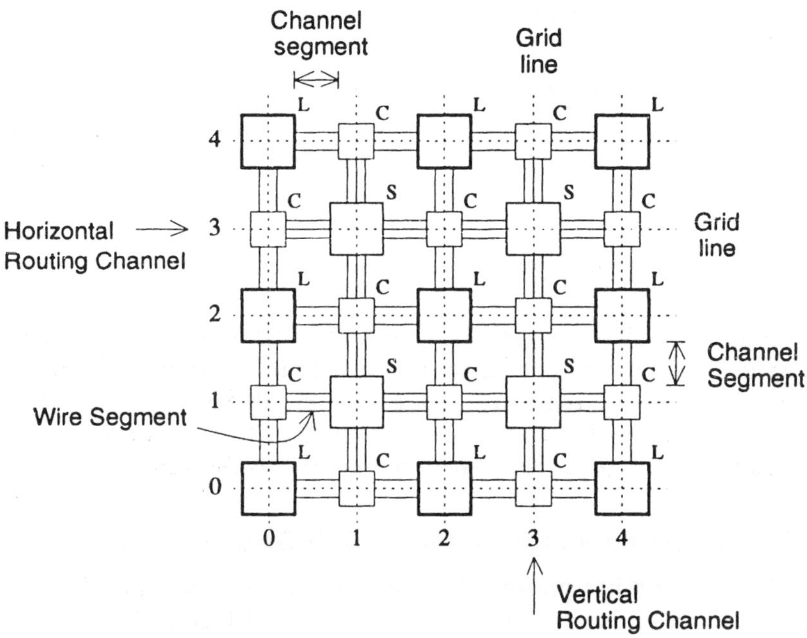

Figure 5.8 - *The Model used for Symmetrical FPGAs.*

Symmetrical FPGAs can be modelled as illustrated by Figure 5.8 [Rose90b] [Rose91]. As the figure shows, the routing channels comprise two kinds of blocks, called Connection (C) blocks and Switch (S) blocks. The C blocks hold routing switches that serve to connect the logic block pins to the wire segments and the S blocks house switches that allow one wire segment to be connected to another. In the figure, each logic block has two pins on each side and there are three tracks in each routing channel. The FPGA's I/O cells appear as logic blocks that are on the periphery of the chip. Figure 5.8 also defines the term *channel segment*, which is a section of a channel between a C block and an S block, or between a C block and a logic block.

The two-dimensional grid that is overlayed on the figure is used later in this section as a means of describing connections.

The general structure depicted in Figure 5.8 is similar to that in Xilinx FPGAs, but it is more general. A wide range of routing architectures can be represented by changing the contents of the C and S blocks. Architectures that feature an abundance of switches would be easily routed, but from the point of view of designing a good routing architecture, the number of switches should be limited, because each switch consumes chip area and has significant capacitance. A routing architecture that has relatively few switches creates difficulties for a routing algorithm. As an example, the following section illustrates the effect on the detailed routing problem if the C blocks allow the logic block pins to connect to only a subset of the wire segments in a channel. This example also serves as the motivation for the detailed routing algorithm that follows.

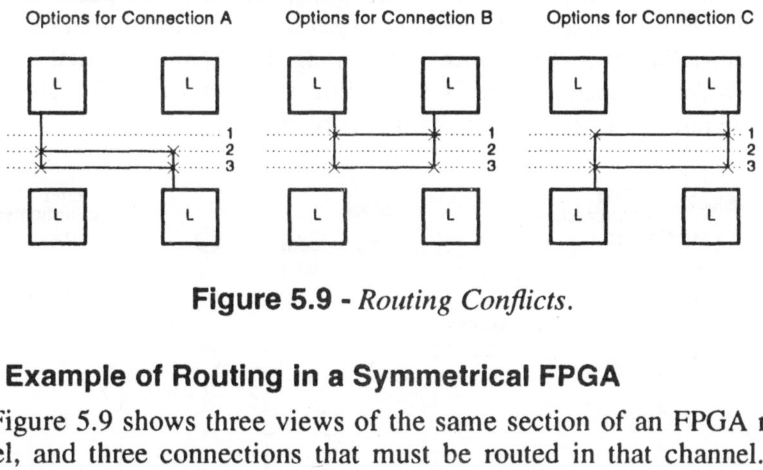

Figure 5.9 - *Routing Conflicts.*

5.4.1 Example of Routing in a Symmetrical FPGA

Figure 5.9 shows three views of the same section of an FPGA routing channel, and three connections that must be routed in that channel. Each view gives the routing options for one of connections A, B, and C. In the figure, a routing switch is shown as an X, a wire segment as a dotted line, and a possible route as a solid line. Now, assume that a detailed router first completes connection A. If the wire segment numbered 3 is chosen for A, then one of connections B and C cannot be routed because they both rely on the same single remaining option, namely the wire segment numbered 1. The correct solution is for the router to choose the wire segment numbered 2 for connection A, in which case both B and C are also routable.

This example shows that even when there are only three connections to be routed, it is possible for a routing decision made for one connection to unnecessarily block another. For this reason, it is important for a detailed routing algorithm for this type of problem to consider the side-effects that routing decisions for one connection have on others. While the example in

Figure 5.9 shows only connections within a single horizontal channel, the problems are compounded when connections have segments that are in both horizontal and vertical channels.

Common approaches used for detailed routing in other technologies are not suitable for symmetrical FPGAs. Maze routers [Lee61] are ineffective because they are inherently sequential, which means that when routing one connection, they cannot consider the side-effects on other connections. Channel routers are not appropriate because the detailed routing problem in symmetrical FPGAs cannot be subdivided into independent channels.

5.4.2 General Approach to Routing in Symmetrical FPGAs

As mentioned earlier in this chapter, the first stage of the routing process is global routing. The global router used is an adaptation of a standard cell global router. It first divides multi-point nets into two-point connections and then chooses routing channels for each one. The main goal of the global router is to distribute the connections among the channels so that the channel densities are balanced. This is a sensible goal for an FPGA global router since the number of tracks per channel is fixed.

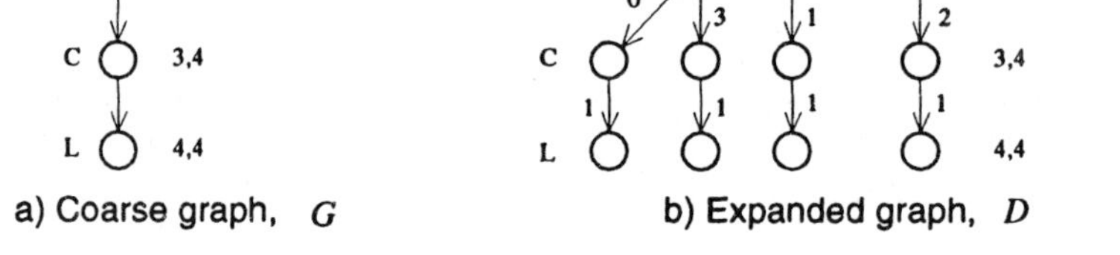

Figure 5.10 - *A Typical Coarse Graph and its Expanded Graph.*

The global router defines a coarse route for each connection by assigning it a sequence of channel segments. Figure 5.10a shows a representation of a typical global route for one connection. It gives a sequence of channel segments that the global router might choose to connect some pin of a logic block at grid location 2,2 to another at 4,4. The global route is called a *coarse graph*, $G(V,A)$, where the logic block at 2,2 is referred to as the root of the graph and the logic block at 4,4 is called the leaf. The vertices, V, and edges, A, of $G(V,A)$ are identified by the grid of Figure 5.8. Since the global router splits all nets into two-point connections, the nodes in the coarse graphs always have a fan-out of one.

After global routing the problem is transformed into the following: for each two-point connection, a detailed router must choose specific wire segments to implement the channel segments assigned during global routing. As this requires complete information about the FPGA routing architecture, the detailed router must use the details of the logic block pins, C blocks, and S blocks to perform its task.

The following section describes a detailed routing algorithm for symmetrical FPGAs. This algorithm, because it accepts the coarse graphs from the global router as input and expands them into detailed routes, is called the *Coarse Graph Expansion* (CGE) detailed router [Brow90] [Brow91]. The algorithm can be used for any FPGA that fits the model shown in Figure 5.8. One of its key features is that it addresses the issue of preventing unnecessary blockage of one connection because of another.

5.4.3 The CGE Detailed Router Algorithm

The basic algorithm is split into two phases. In the first phase, it records a number of alternatives for the detailed route of each coarse graph, and then in the second phase, viewing all the alternatives at once, it makes specific choices for each connection. The decisions made in phase 2 are driven by a cost function that is based on the alternatives enumerated in phase 1. Multiple iterations of the two phases are used to allow the algorithm to conserve memory and run-time while converging to its final result, as discussed in Section 5.4.3.3.1.

5.4.3.1 Phase 1: The Expansion of the Coarse Graphs

During phase 1, CGE *expands* each coarse graph and records a subset of the possible ways that the connection can be implemented. For each $G(V,A)$, the expansion phase produces an *expanded* graph, called $D(N,E)$. N are the vertices of D and E are its edges, with each edge referring to a specific wire segment in the FPGA. The edges are labelled with a number that refers to the corresponding wire segment.

In the expansion algorithm, the procedures that define the connection topology of the C and S blocks are treated as *black-box* functions. The black-box function for a C block is denoted as $f_c([d_1,d_2,l],d_3)$ and for an S block as $f_s([d_1,d_2,l],d_3)$. The parameters in square brackets define an edge that connects vertex d_1 to vertex d_2, using a wire segment labelled l. Such an edge is later referred to as e, where $e = (d_1,d_2,l)$. The parameter d_3 is the successor vertex of d_2 in G. The task of the function call can be stated as: "If the wire segment numbered l is used to connect vertex d_1 to d_2, what are the wire segments that can be used to reach d_3 from d_2?" The function call returns the set of edges that answer this question. As explained in Section 5.4.3.4, this black-box approach provides independence from any specific FPGA routing architecture. The result of a graph expansion is illustrated in Figure 5.10b, which shows a possible expanded graph for the coarse graph of Figure 5.10a. An expanded graph is produced by examining the routing switches and wire segments along the path described by the coarse graph, and recording the alternative detailed routes in the expanded graph. In algorithmic form, the graph expansion process for each coarse graph operates as follows:

> **create** D and give it the same root as G. Make the immediate successor to the root of D the same as for the root of G
> **for** each new vertex, traversing D breadth first {
>   **expand** a C vertex in D by calling $Z = f_c(e_C,n)$. e_C is the edge in D that connects to C from its predecessor. n is the required successor vertex of C (in G) and Z is the set of edges returned by $f_c(\,)$. The call to $f_c(\,)$ adds Z to D
>   **expand** an S vertex in D by calling $Z = f_s(e_S,n)$. e_S is the edge in D that connects to S from its predecessor. n is the required successor vertex of S (in G) and Z is the set of edges returned by $f_s(\,)$. The call to $f_s(\,)$ adds Z to D
> }

5.4.3.2 Phase 2: Connection Formation

After expansion, each $D(N,E)$ may contain a number of alternative paths. CGE places all the paths from all the expanded graphs into a single path list. Based on a cost function, the router then selects paths from the list; each selected path defines the detailed route of its corresponding connection. Phase 2 proceeds as follows (as explained later in this section, the terms c_f cost and c_t cost are functions that represents the relative cost of selecting a specific detailed route (path) for a connection, and an *essential* path indicates a connection that should be routed immediately because it has only one remaining option):

> **put** all the paths in the expanded graphs into the path-list
> **while** the path-list is not empty {
>   **if** there are paths in the path-list that are known to be essential
>   **select** the essential path that has the lowest c_f cost

else if there are paths in the path-list that correspond to time-critical connections
select the critical path with the lowest c_t cost
else
select the path with the lowest c_f cost
mark the graph corresponding to the selected path as routed - remove all paths in this graph from the path-list
find all paths that would conflict with the selected path and remove them from the path list (see Note). If a connection loses all of its alternative paths, re-expand its coarse graph - if this results in no new paths, the connection is deemed unroutable (see Section 5.4.3.3.1 for a discussion relating to failed connections).
update the cost of all affected paths
}

Note: When a wire segment is chosen for a particular connection, it and any other wire segments in the FPGA that are hardwired to it must be eliminated as possible choices for connections that are in other nets. This requires a function analogous to $f_c()$ and $f_s()$ that understands the connectivity of a particular FPGA configuration. CGE calls this routine *update* (e) - the parameter e is an edge in the selected path and *update* (e) returns the set of edges that are hardwired to e.

5.4.3.2.1 Cost Function

Because the cost function allows it to consider all the paths at once, CGE can be said to route the connections 'in parallel'. Each edge in the expanded graphs has a two-part cost: $c_f(e)$ accounts for the competition between different nets for the same wire segments, and $c_t(e)$ is a number that reflects the routing delay associated with the wire segment. Each path has a cost that is simply the sum of the costs of its edges. CGE selects paths based on the c_t cost only if the path corresponds to a time-critical connection. Otherwise, paths are selected according to their c_f cost.

The c_f cost has two goals:

1. To select a path that has a relatively small negative effect on the remaining connections, in terms of routability. The cost deters the selection of paths that contain wire segments that are in great demand. The reason for using wire segment demand was illustrated in Figure 5.9, where connection A should be routed with wire segment number 2, because wire segment number 3 is in greater demand.

2. It is used to identify a path that is *essential* for a connection. A path is called essential when it represents the only remaining option in the FPGA for a connection, because previous path selections have consumed all other alternatives.

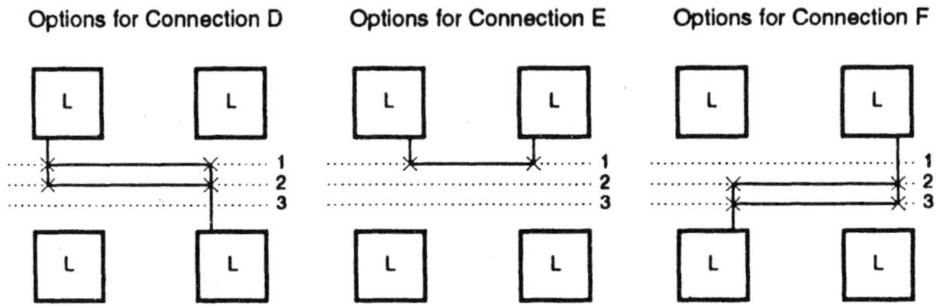

Figure 5.11 - *An Essential Wire Segment.*

The importance of essential wire segments is illustrated by the example in Figure 5.11. If the router were to complete connection D first, then wire segment number 1 or 2 would be equal candidates according to their demand, since they both appear in one other graph. However, wire segment number 1 is essential for the completion of connection E and to ensure the correct assignment of the essential wire segment, connection E should be routed first.

To determine whether an edge, e, is in great demand the router could simply count the number of occurrences of e that are in expanded graphs of other nets. However, some occurrences of e are less likely to be used than others because there may be alternatives (edges in parallel with e). Thus, the c_f cost of an edge e that has j other occurrences ($e_1, e_2, ..., e_j$) is defined as

$$c_f(e) = \sum_j \frac{1}{alt(e_j)},$$

where $alt(e_j)$ is the number of edges in parallel with e_j.

Because of the summing process in $c_f(e)$, the more graphs e occurs in, the higher will be its cost. This reflects the fact that e is an edge that is in high demand and urges CGE to avoid using e when there are other choices. Note that an edge that only appears in its own graph will have a c_f of 0. For the special case when $alt(e_j)$ is 0, e_j is an edge that is essential to the associated connection because there are no alternatives. In this case, any path in the graph that uses e_j is identified as *essential*. When the calculation of a cost reveals that a path is essential, CGE gives that path the highest priority for routing.

5.4.3.3 Controlling Complexity

Although the above description of graph expansion implies that all possible paths in an FPGA are recorded during expansion, this is not practical because the number of paths can be very large in some architectures. For example, consider the connection of two pins on two different L blocks. Assume that each pin can connect to F_c of the wire segments in the channel segments adjacent to each logic block, and that the logic blocks are separated by n Switch blocks. If each wire segment that enters one side of a Switch block can connect to F_s wire segments on the other three sides, then there are an average of $F_c(F_s/3)^n$ different paths from the first pin to the last logic block, and assuming W tracks in each routing channel, there are an average of $(F_c^2/W)(F_s/3)^n$ possible ways to form the connection. Research has shown [Rose90b] [Rose91] that typical values of F_s should be three or greater, and since the number of connections is large, a heuristic is employed to reduce the number of paths in the expanded graphs. Some of the paths are pruned as each graph is expanded. The pruning procedure is parameterized so that the number of paths is controlled and yet the expanded graphs still contain as many alternatives as possible. Maximizing the number of alternatives is important in the context of resolving routing conflicts. The pruning procedure is part of the graph expansion process that was described in Section 5.4.3.1. The general flow follows (the criteria used for pruning is given at the end of this section):

> **expand** two levels
> **prune**; keep at most K vertices at this level, and assign each a unique group number. Discard the other vertices and the paths they terminate
> **expand** two more levels. Assign each added vertex the group number of its predecessor
> **while** the leaf level has not been reached {
>   **prune**; keep at most k vertices with each group number at this level. Discard the other vertices and the paths they terminate
>   **expand** two more levels. Assign each added vertex the group number of its predecessor
> }

The graphs are pruned every two levels because that is where fanout occurs (after the first C block and after every S block). The parameter K controls the starting widths of the graphs and can take values from one to F_c (the number of wire segments connected to each logic block pin). Beyond the maximum value of K, parameter k allows the expanded graphs to further increase in width. The concept of *group* numbers isolates each of the original K paths, which maximizes the number of alternatives at each level of the final expanded graph. The actual values used for K and k are discussed in the next section. The effect of the pruning algorithm is illustrated in Figure 5.12. The left half of the figure shows a fully expanded graph from an example circuit, while the corresponding pruned graph is on the right. Also shown are each graph's edges in the FPGA.

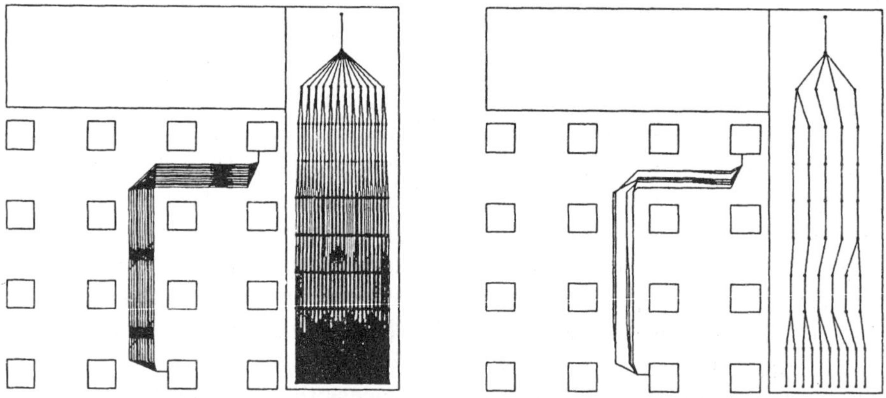

Figure 5.12 - *The Effect of Pruning.*

The choice to prune a vertex is based on the wire segment that corresponds to its incoming edge, as follows. For the special case of time-critical connections, the wire segments with the least delay are favored. For other connections, the wire segments that have thus far been included in the most other expanded graphs will be discarded. This helps the c_f cost function discover the wire segments that are in the least demand. Note that this introduces an order-dependence in the routing algorithm because the paths that are pruned from each expanded graph depend on the order in which the coarse graphs are expanded.

Note that when paths are discarded because of pruning, they are not necessarily abandoned permanently by the router. In phase 2, as CGE chooses connections, if routing conflicts consume all the alternatives for some graph, CGE re-invokes the graph expansion process to obtain a new set of paths if some exist.

5.4.3.3.1 Iterative Improvements

This section explains how iterations of the two phases of CGE are used to conserve memory and run-time. The iterative approach is linked to the pruning parameters of the graph expansion phase. Setting the pruning parameters to large values allows the router to do a better job of resolving routing conflicts because it sees many alternatives for each connection. On the other hand, with large pruning parameters more memory and longer run-time are required by the algorithm. The key to this routing quality versus memory and time trade-off is the realization that most connections in an FPGA are relatively easy to route and only a small percentage of the connections pose real difficulties. This is because, in a typical routing problem, there are only a few channel segments whose densities are very close to the total number of wires in a routing channel. To exploit this property, the router starts with small pruning parameters and then increases them through successive iterations, but only for the parts of the FPGA that are difficult to route.

For the first iteration the pruning parameters are set to relatively small values, and the entire FPGA is routed. If routing conflicts leave some connections unrouted, then another iteration is required. The procedure is to erase all the routing of any connection that overlaps any part of a failed connection, and then to attempt to route those channel segments again using larger pruning parameters. Only connections that touch some segment of a channel in which a failed connection occurred are re-routed in the next iteration. Iterations are continued until all connections are routed or until further improvements are not forthcoming.

At this point it would be desirable to try different global routes for connections that are left unrouted after all iterations, but no such failure-recovery mechanism is currently implemented. This iterative approach is a minor variation of classic rip-up and re-route schemes where individual connections would be removed and re-routed to try to resolve routing conflicts. The technique employed here allows the algorithm's cost function to solve the routing problem, but conserve memory and time where the problem is not difficult and expend them only where it is required.

The specific values used for the pruning parameters in each iteration affect the total number of iterations required, but do not appreciably affect the quality of the final result. This indicates a robustness in the algorithm, because the quality of the routing does not depend on the specific values chosen for the program's parameters. For the results that are presented in Section 5.4.3.5, K and k are set to two for the first iteration. K is increased by one for each iteration until it reaches F_c, after which k is increased by one for each subsequent iteration.

5.4.3.4 Independence from FPGA Routing Architectures

CGE achieves the ability to route arbitrary FPGA routing architectures by isolating the parts of the code that are architecture-specific. This is illustrated in Figure 5.13 which shows the overall flow of the algorithm. The code that is dependent on the routing architecture is enclosed in circles. As shown, the separate code includes the f_c (), f_s (), and *update* () routines. Any architecture that fits the general model in Figure 5.8 can be routed by changing these isolated routines. This generality has been utilized in a study of FPGA routing architectures that is described in Chapter 6 of this book. Figure 5.13 also shows the organization of the phases of CGE and the feedback path used over multiple iterations.

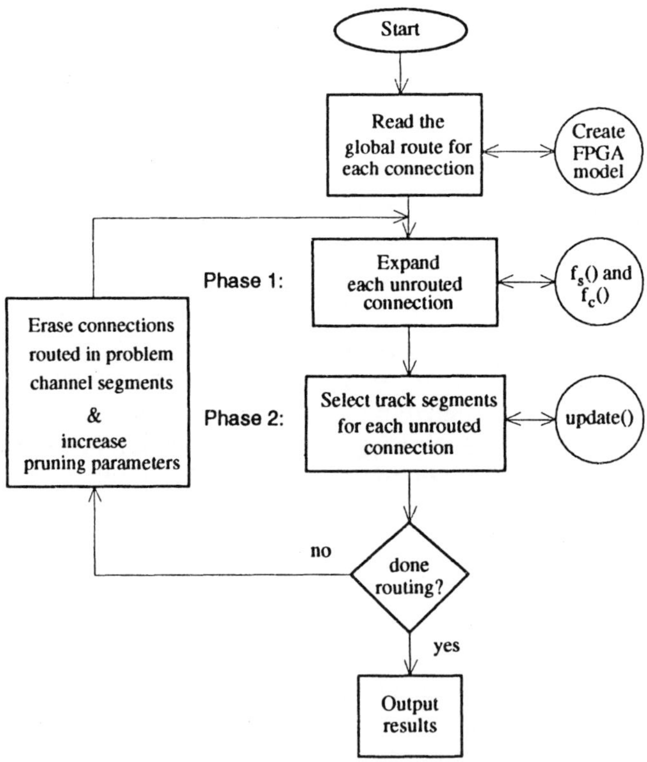

Figure 5.13 - *The Organization of CGE.*

5.4.3.5 Results

This section presents the results of using CGE to route several industrial circuits in symmetrical FPGAs. The routing results shown in this section are based on five circuits from four sources: Bell-Northern Research, Zymos, and two different designers at the University of Toronto. Table 5.1 gives the name, size (number of two-point connections and logic blocks), source and the function of each circuit. For these results, the logic block used is the result of a previous study [Rose89] [Rose90c], and the S and C blocks will be described in the next section. For these results, the C and S blocks are defined so that the routing architecture is quite similar to that in the Xilinx 3000 series FPGAs that were described in Chapter 2. The similarity refers to the amount of connectivity that is available between the logic block pins and the wire segments and between one wire segment and another.

Circuit	#Blocks	#Conn	Source	Type
BUSC	109	392	UTD1	Bus Cntl
DMA	224	771	UTD2	DMA Cntl
BNRE	362	1257	BNR	Logic/Data
DFSM	401	1422	UTD1	State Mach.
Z03	586	2135	Zymos	8-bit Mult

Table 5.1 - *Experimental Circuits.*

5.4.3.5.1 FPGA Routing Structures

Since the routability of an FPGA is determined by the topology and flexibility of its S and C blocks, those used in the tests of the algorithm are presented here. The general nature of the S block is illustrated in Figure 5.14a. Its flexibility is set by the parameter F_s, which defines the total number of connections offered to each wire segment that enters the S block. For the example shown in Figure 5.14a the wire segment at the top left of the S block can connect to six other wire segments, and so F_s is 6. Although not shown, the other wire segments are similarly connected.

Figure 5.14b illustrates the test C block. The tracks pass uninterrupted through it and are connected to logic block pins via a set of switches. The flexibility of the C block, F_c, is defined as the number of tracks that each logic block pin can connect to. For the example shown in the figure, each logic block pin can connect to 2 vertical tracks, and so F_c is 2.

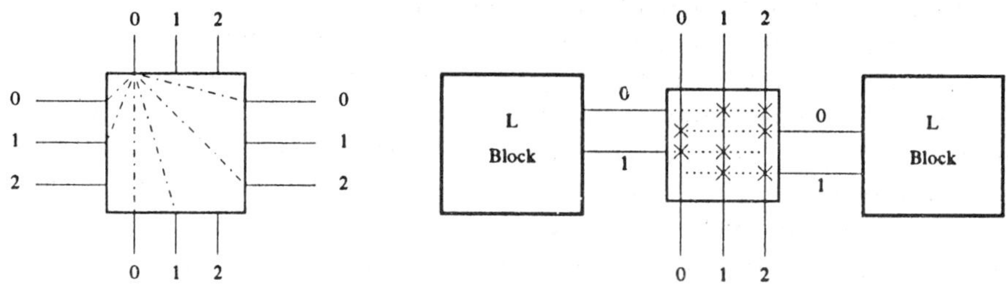

Figure 5.14 - *Definitions of S and C Block Flexibility.*

5.4.3.5.2 Routing Results

The familiar yardstick of channel density is used as a measure of the quality of the detailed router. The 'Channel density' column in Table 5.2 shows the maximum channel density over all channels for each circuit. This represents a lower bound on the number of tracks per routing channel that is needed for each example. The real track requirements will depend on the flexibility of the routing architecture, because the channel density measure does not consider the amount of connectivity that is available in the routing structures. The maximum flexibility has $F_s = 3W$ and $F_c = W$, where there are W tracks per channel. For the results in Table 5.2, the FPGA parameters are based on the Xilinx 3000 series [Xili89] ($F_s = 6$, $F_c = 0.6W$). Table 5.2 gives the minimum number of tracks per channel that CGE needs in order to route 100 percent of the connections. The values for W are slightly greater than the global router minimum, which are excellent results considering the low flexibility of the FPGA routing architecture. Note that, although not shown, if F_c is increased to 0.8W, CGE achieves the absolute minimum number of tracks for all the circuits.

For comparison purposes, the same problems have also been routed using CGE with its c_f cost facility disabled. In this mode CGE has no ability to resolve routing conflicts and is thus a sequential router, similar to a maze router. At first glance, this may seem to be an unrealistic comparison because some maze routers are guided by cost functions that aid in finding good routes for connections. However, the 'maze' router used here has, in effect, access to the cost function that was used to solve the global routing, which is based on balancing the densities of all routing channels. Notwithstanding, this is a constrained 'maze' router because it is confined to remain within the global route of each connection, and the comparisons are valid

Circuit	Channel density	W required by CGE	W for 'maze'
BUSC	9	10	15
DMA	10	10	15
BNRE	11	12	20
DFSM	10	10	18
Z03	11	13	18

Table 5.2 - *CGE Minimum W for 100 % routing ($F_c = 0.6W$, $F_s = 6$).*

only in that context. The rightmost column in Table 5.2 gives the number of tracks that the 'maze' router requires to achieve 100 percent routing. These results demonstrate that the 'maze' router needs an average of 60 percent more tracks than CGE. This shows that resolving routing conflicts is important and that CGE addresses this issue well. Figure 5.15 presents the detailed routing for circuit BUSC, with the FPGA parameters in Table 5.2; the logic blocks are shown as solid boxes, whereas the S and C blocks are dashed boxes.

Name of net	# of switches without critical processing	# of switches with critical processing
#143	15	5
#144	14	4
#220	10	3
#280	15	2
#351	15	4

Table 5.3 - *Critical Connection Routing Delay Optimization.*

5.4.3.5.3 Routing Delay Optimization for Critical Nets

Table 5.3 illustrates CGE's ability to optimize critical connections. For this experiment, several connections in circuit BNRE were marked critical. Then, CGE was used to route the circuit twice; once with CGE's critical net processing turned off, and once with it turned on. To facilitate this experiment, the FPGA was defined to have 18 tracks per channel, with four tracks hardwired for the entire length of each channel. Connections that use the hardwired tracks have lower routing delays because they pass through fewer switches (transistors). As Table 5.3 shows, a significant reduction in the number of switches in the critical paths was achieved.

Note that a better approach to routing delay optimization would set specific timing requirements that should be met for each critical *path* in a circuit. This possibility should be explored.

5.4.3.5.4 Memory Requirements and Speed of CGE

For the examples used here CGE needs between 1.5 and 7.5 MBytes of memory. As shown in Table 5.4, experimental measurements show that CGE is a linear-time algorithm, requiring from 25 to 215 SUN 3/60 CPU seconds for the smallest to the largest of the example circuits. This run-time behavior is due to the pruning procedure, which limits the number of routing alternatives that the algorithm considers for each connection.

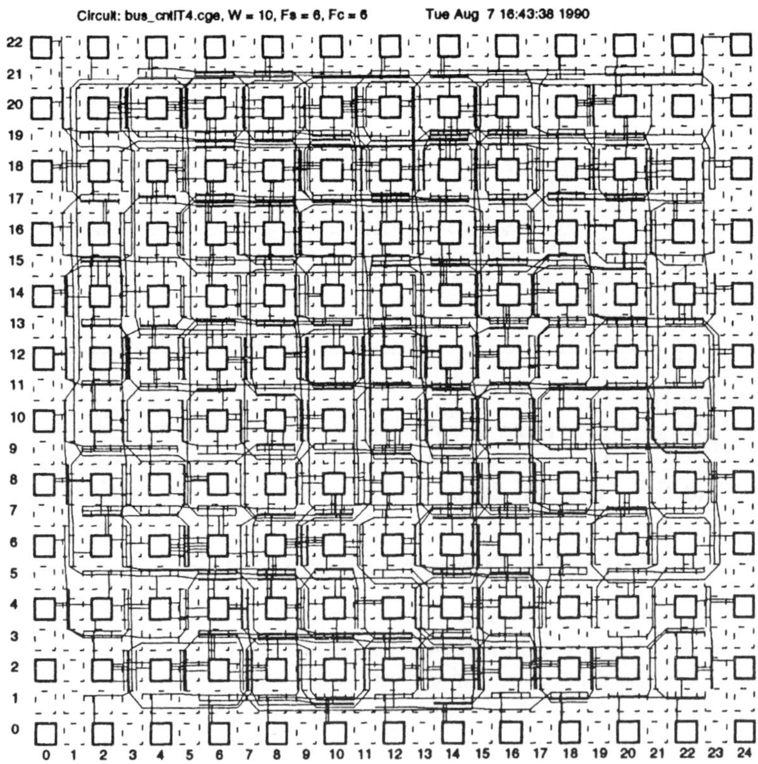

Figure 5.15 - *The Detailed Routing of Circuit BUSC.*

Circuit	#Conn	Sun 3/60 CPU sec.	msec per connection
BUSC	392	25	63
DMA	771	59	76
BNRE	1257	122	97
DFSM	1422	103	72
Z03	2135	215	99

Table 5.4 - *CGE Run-time.*

5.4.4 Final Remarks for Symmetrical FPGAs

This section has discussed some of the important issues for routing in symmetrical FPGAs. A detailed routing algorithm has been described that is able to consider the side-effects that routing decisions made for one connection may have on another, which is important in an FPGA that has limited connectivity. The routing algorithm can be used over a wide range of FPGA routing architectures and can route relatively large FPGAs using close to the absolute minimum number of tracks as determined by global routing.

Since the CGE detailed router can handle a wide range of routing architectures, it can be employed to investigate the effects of an FPGA's routing structures on the routability of circuits. Such a study has been conducted and is the subject of the next chapter.

CHAPTER 6

Flexibility of FPGA Routing Architectures

Chapter 5 discussed the important issues associated with designing good CAD tools for routing in FPGAs. This chapter focuses on a related issue, that of designing FPGA routing architectures. The reader will recall that the results of some routing architecture experiments were already presented, in Chapter 4. Those results concerned the segmentation of routing channels in row-based FPGAs and showed that it is important to tune the segmentation of the channels for the type of K-segment routing that is desired (see Section 5.3.5). In this chapter, more detailed results are presented from another study of routing architectures, this one concerning symmetrical FPGAs [Rose90b] [Rose91]. The approach taken is an experimental one, in which circuits are implemented by CAD tools in symmetrical FPGAs, over a range of different routing architectures. The main CAD tool used for the experimental results is the CGE detailed router that was described in Section 5.4.3.

FPGA routing architectures can be designed in many ways. Some of the design parameters are the number of tracks per routing channel, the connectivity available between the logic block pins and the wire segments, and the connectivity between one wire segment and others. The *routing architecture* in an FPGA includes all of its routing switches and wire segments, and their distribution over the surface of the chip. A measure of the connectivity provided by a routing architecture is its *flexibility*, which is a function of the total number of routing switches and wires.

Designing a good routing architecture involves a tradeoff among flexibility, logic density, and speed performance. An FPGA with a high flexibility will be easy to configure, but if the flexibility is too high then area will be wasted by unused switches, leaving less area for the logic blocks and resulting in lower logic density. Moreover, as was shown in Chapter 2, since each routing switch introduces an RC-delay, high flexibility results in reduced speed performance. Low flexibility, on the other hand, allows higher logic density and lower RC-delay, but if the flexibility is too low, then it may not be possible to interconnect the logic blocks sufficiently to implement circuits. The study that is presented in this chapter investigates these competing factors, toward the goal of finding routing architectures that achieve a balance.

6.1 FPGA Architectural Assumptions

The routing architectures that are studied in this chapter fit the symmetrical FPGA model that was introduced in Section 5.4. Figure 6.1 reproduces the model, for ease of reference. Recall that the connection (C) blocks contain routing switches that are used to connect the pins of the logic blocks to the routing channels, and the switch (S) blocks at the intersections of horizontal and vertical routing channels provide the routing switches that can connect wire segments in one channel segment to those in another. The flexibility of the routing architecture can be altered by changing the contents of the C blocks, the S blocks or the number of tracks in each routing channel.

In this chapter, circuits are implemented in this model of the FPGA, with different numbers of routing switches in the C and S blocks, and with a range of tracks per routing channel. The experiments answer a number of specific questions concerning the effect of the routing structures on the implementation of circuits:

- What is the effect of the flexibility of the C blocks on routability?
- What is the effect of the flexibility of the S blocks on routability?
- How do the S block and C block flexibilities interact?
- What is the effect of the flexibilities of the C and S blocks on the number of tracks per routing channel required to achieve 100 percent routability?
- What is the effect of the flexibilities of the C and S blocks on the total number of routing switches required in an FPGA to achieve 100 percent routability?

The FPGA model shown in Figure 6.1 is quite general, allowing for a range of different logic blocks and routing structures. The following sections detail the specific assumptions that are made for the FPGA's architecture for the experimental results that are presented later in this chapter.

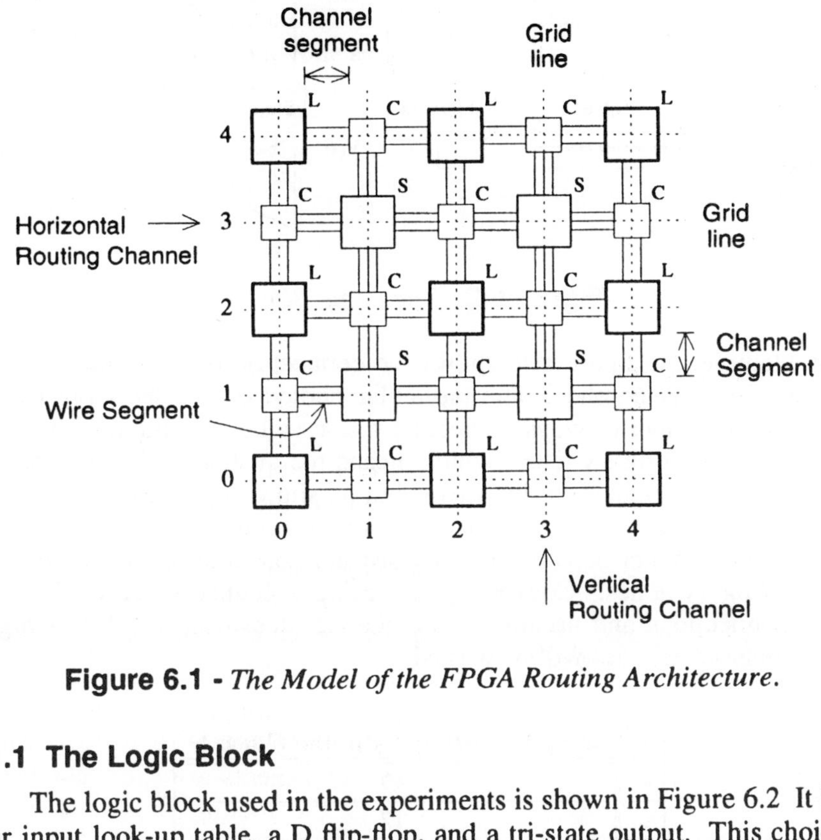

Figure 6.1 - *The Model of the FPGA Routing Architecture.*

6.1.1 The Logic Block

The logic block used in the experiments is shown in Figure 6.2 It has a four input look-up table, a D flip-flop, and a tri-state output. This choice of logic block was based on the study [Rose89][Rose90c] of logic block architectures that described in Chapter 4. As Section 4.2.2 showed, the logic block in Figure 6.2 resulted in the minimum total chip area when compared to other blocks that had differing numbers of inputs, including and excluding a D flip-flop. The logic block has a total of 7 logical pins (numbered from 0 to 6 in the figure, for later reference): 4 inputs, 1 output, 1 clock, 1 tri-state enable. Each pin may physically appear on one or more sides of the block. As explained below, the number of physical occurrences of each pin is an important architectural parameter.

The number of logic block sides on which each logical pin physically appears is called T. To illustrate this concept, the cases $T=4$ and $T=1$ are shown in Figure 6.3, where the logic block pins are numbered from 0 to 6.

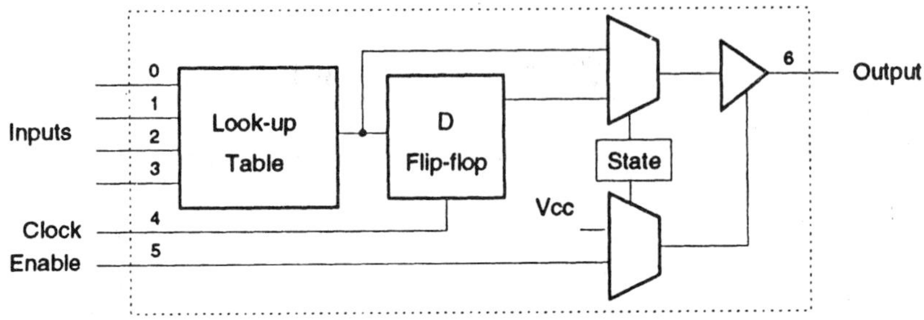

Figure 6.2 - *The Logic Block.*

As an example, the figure also shows the connection of pin 0 on one logic block to pin 6 on another. The particular choice of T affects the routing problem in a number of ways. Selecting a low value of T implies that there will be fewer routing switches, which means the switches will use less area and add less capacitance to the tracks, but as shown in Figure 6.3, connections may be longer since it may be necessary to route to a certain side. This increases the channel densities and causes the connections to pass through more routing switches. Conversely, choosing a higher value of T allows shorter connections and minimizes the channel densities, but if T is higher than necessary, switches will be wasted.

T	Avg. Maximum Channel Density
1	15
2	12
3	11
4	11

Table 6.1 - *The Effect of T on Channel Density.*

For the experiments shown here, the value used for T is 2. This was chosen for area considerations only and was determined by performing the global routing of several circuits for each value of T and measuring the number of tracks per channel (maximum channel density) required in each case. The results are shown in Table 6.1, which gives the average maximum channel density of the circuits, for each value of T. The table shows a significant decrease in track count from the $T=1$ to $T=2$ case but diminishing returns for higher values of T. Note that the routing tools used for these experiments did not make use of the functional equivalence of the logic block inputs (the inputs to a look-up table are functionally equivalent), and if they had it would have been possible to choose a value of $T=1$, without an increase in the number of tracks [Tseng92].

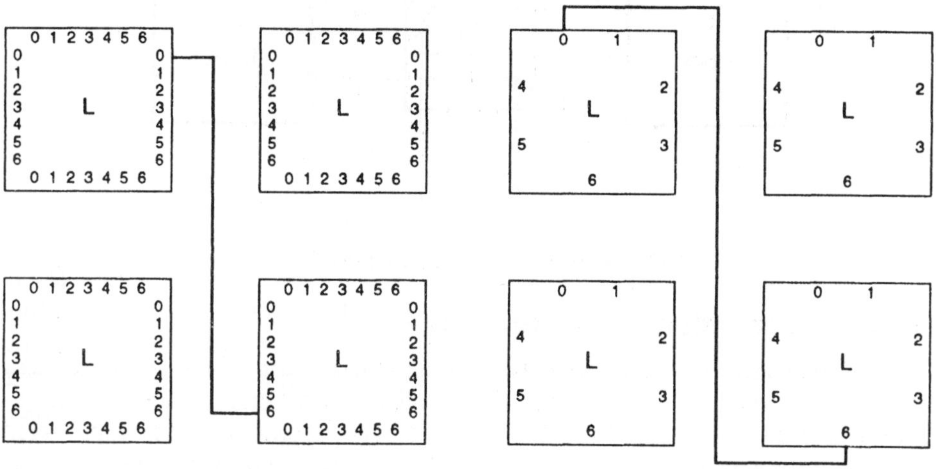

T = 4
Short connection is possible

T = 1
Longer connection is necessary

Figure 6.3 - *Example, and Effect on Connection Length, of T.*

The following two sections provide a detailed discussion of the C and S blocks used in the experiments. Some of this information also appeared in Chapter 5, but is repeated here for continuity.

6.1.2 The Connection Block

The connection block used is illustrated in Figure 6.4, where a routing switch is indicated by an X. The channel wires (drawn vertically in the figure) pass uninterrupted through the C block and have the option of connecting to the logic block pins through the switches. The flexibility of the C block is represented by the variable F_c, which defines the number of tracks that each logic block pin can connect to. For the example shown in the figure, each logic block pin can connect to 2 vertical tracks, and so F_c is 2. In this chapter, no assumption is necessary about the implementation of routing switches, except that the switches are assumed to be bi-directional.

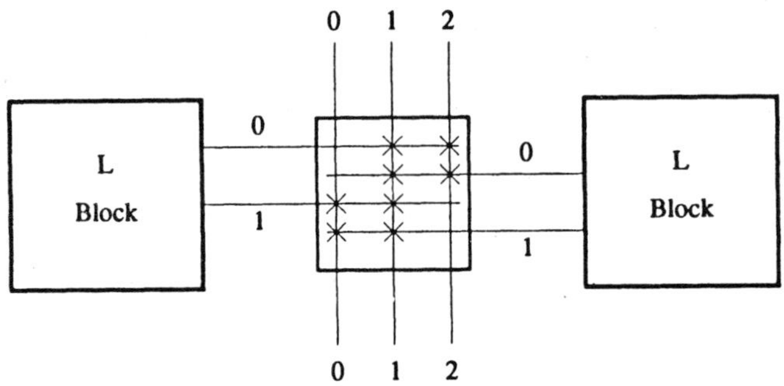

Figure 6.4 - *The Connection Block.*

6.1.2.1 Connection Block Topology

The *topology* of the connection block (the pattern of the switches) can have a significant effect on routability, particularly when F_c is low. To illustrate this consider Figure 6.5 which shows two different C block topologies and one connection (from pin A to pin B) that must be routed. In this example each logic block has three pins on a side and there are four tracks per routing channel. By examining the locations of the routing switches, it is clear that it is not possible to route the connection with Topology 1, while it is possible to do so using Topology 2. Topology 2 works because it has a common wire that can be reached by both pin A and pin B. This example illustrates the fact that a C block topology must provide common wires for every pair of pins that may need to be connected. At the same time, however, it is easy to recognize that it is desirable for the routing switches in the C block to be spread evenly among the tracks, so that there is a reasonable opportunity for each track to be used. A good C block topology should achieve a balance of these tradeoffs. Given W tracks per channel, the design of a C block is straight-forward if F_c is close to W, but for lower values of F_c the C block should be carefully designed. The issue is most acute if $F_c \leq 0.5W$, because at this point some pairs of pins may not have any common wires if the C block is poorly designed.

The topology of the C block that is used for the results presented in this chapter is illustrated by Figure 6.6. In the figure, there are 10 tracks per routing channel, seven pins per logic block, and F_c is 6. The design of this topology is based on statistics, from a set of circuits, that show how frequently each pair of pins is connected. For pins that are often connected the topology tends to provide common tracks, whereas for pins that are seldom connected different tracks are used. As an example, the statistics say that pin 0 (a logic

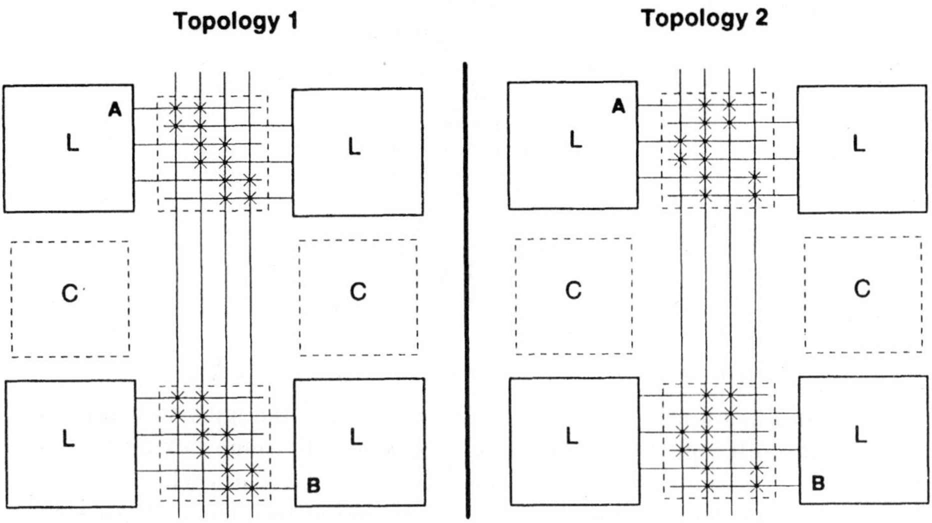

Figure 6.5 - *Two Connection Block Topologies.*

block input, as shown in Figure 6.2) is often connected to pin 6 (an output), so these two pins share six tracks, whereas pin 0 is seldom connected to pin 5 (an input), so this pair shares only three tracks. This type of analysis is possible because logic block inputs tend to be connected to outputs, and vice-versa. In this way, the topology provides as much overlap as practical for each pair of logic block pins, while also balancing the distribution of the switches among the channel wires.

6.1.3 The Switch Block

The general nature of the switch block used is illustrated in Figure 6.7. Its flexibility, F_s, defines the number of other wire segments that each wire segment entering an S block can connect to. For the example shown in the figure, the wire segment at the top left of the S block can be switched to six other wire segments, and so F_s is 6. Although not shown, the other wire segments are similarly connected.

6.1.3.1 Switch Block Topology

The topology of the S blocks can be very important since it is possible to choose two different topologies with the same flexibility measure (F_s) that result in very different routabilities. This is particularly important if the flexibility is low.

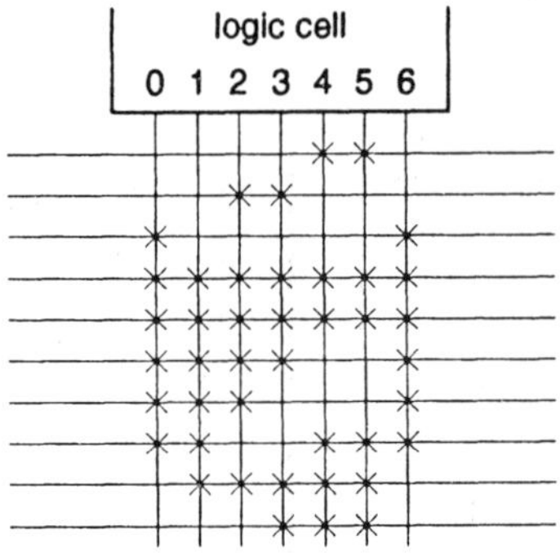

Figure 6.6 - *The Experimental C Block Topology.*

As an illustration, consider the two different topologies shown in Figure 6.8. In both topologies, the switch block has the same flexibility measure, $F_s = 2$. Assume that a global router has specified that a wire segment at A must be connected to another at B by traveling through the two switch blocks shown. By examining the routing switches, it is easy to see that it is not possible to reach B from A with topology 1, while it is with topology 2. The reason that topology 2 is successful can be explained as follows. Consider the two vertical wires in topology 2 that connect from A to the two horizontal wires on the right side of the S block. At the next S block, one of the horizontal wires can connect to the top of the block (to B) and one to the bottom. The key is that any turn that is taken at one S block does not prohibit any other turn at the next S block, and this is true for all possible sequences

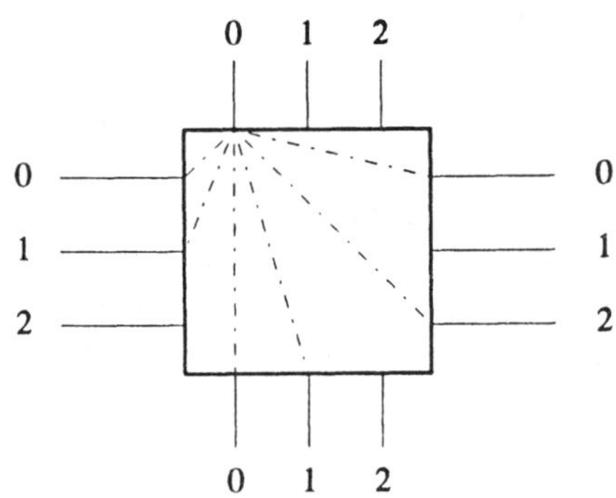

Figure 6.7 - *The Switch Block.*

of turns. For the results that are presented in this chapter, topology 2 is used. For higher flexibilities, switches are added such that the basic pattern is preserved.

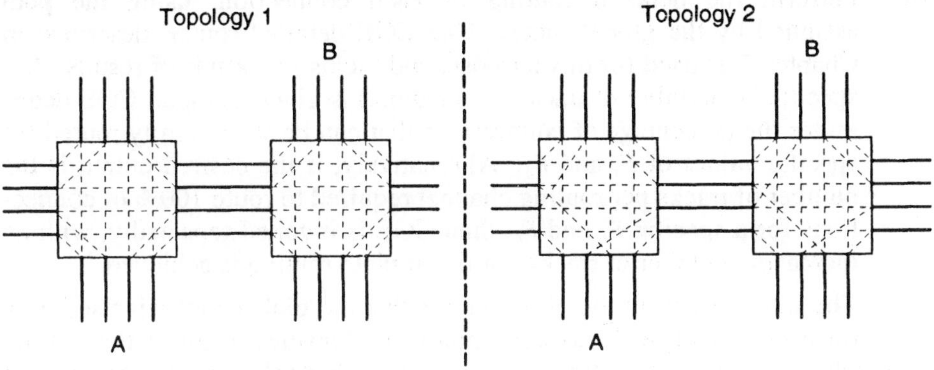

Figure 6.8 - *Two Switch Block Topologies.*

6.2 Experimental Procedure

This section describes the experimental procedure that is used to investigate FPGA routing architectures [Rose90b] [Rose91]. Given a functional description of a circuit, the experimental procedure is as follows:

(1) Perform the *technology mapping* [Keut87] [Fran90] [Fran91a] of the original network into the FPGA logic blocks. As discussed in Chapter 3, this step transforms the functional description of the network into a circuit that interconnects only logic blocks of the type shown in Figure 6.2. The technology mapping was done using an early version of the Chortle algorithm, which was described in Chapter 3.

(2) Perform the placement of the netlist of logic blocks. The logic blocks were placed by the Altor placement program [Rose85], which is based on the min-cut placement algorithm [Breu77]. Altor makes the resulting two-dimensional array of logic blocks as square as possible.

(3) Perform the global routing of the logic block interconnections. As discussed in Chapter 5, this step finds a path through the routing channels for each pair of logic block pins that are to be connected. Since each connection is assigned to specific channels this determines the *maximum channel density* of the circuit, which is defined as the maximum number of connections that pass through any channel segment. This sets the theoretical minimum number of tracks per channel (for the particular global router used) that is needed to route the circuit. The

global router employed for the results presented here is based on the LocusRoute standard cell global routing algorithm [Rose90a].

(4) Perform the detailed routing of each connection, using the path assigned by the global router. The CGE detailed router, described in Chapter 5, is used for this purpose, and yields two kinds of results. If a specific W (number of tracks per channel) is given as input, CGE determines the percentage of connections that can be successfully routed for specific values of F_s and F_c. Alternatively, if the desired output is the number of tracks per routing channel required to route 100% of connections for a specific F_s and F_c, then CGE is invoked repeatedly, with an increasing number of tracks, until complete routing is achieved.

The salient point in this procedure is that the global router is used only once for each circuit, and this determines the densities of all of the routing channels. The number of tracks required per channel to route each circuit then depends on the flexibility of the routing architecture. Thus, to investigate the effect of flexibility on routability, step (4) was performed over a range of values of F_c, F_s, and W.

6.3 Limitations of the Study

This section discusses the effects of the architectural assumptions and the experimental procedure on the accuracy of the results that are presented later in this chapter.

The models that have been used for the C and S blocks are based on balanced topologies, in that each L block pin can be connected to exactly F_c tracks and each wire segment that enters an S block can connect to exactly F_s others. Also, every wiring track must use a routing switch to pass through an S block - i.e. all the tracks comprise short wire segments only. Although it is also interesting to consider other classes of architectures, the assumptions made here allow interesting and useful results to be generated with experiments that have simple parameters. A study of track segmentations for symmetrical FPGAs, such as that described in Section 5.3.5 for row-based FPGAs, should be the subject of future research.

The experimental procedure described in Section 6.2 limits each connection to a single global route. A better approach would be one that provides a feedback mechanism that allows the detailed router to request a different global route for connections that fail. Finally, the accuracy of the routability results that are presented in this chapter depends on the quality of the routing CAD tools, which includes both the global and detailed routers.

6.4 Experimental Results

The experimental results that are presented here are based on the five circuits that were described in Table 5.1. This section first investigates the effect of the flexibilities of the C and S blocks on the routability of these circuits and shows the tradeoffs that exist between these two blocks. Following this, the effect of different values of F_c and F_s on the number of tracks required per channel is shown. Finally, the effect of the C and S block flexibilities on the total number of switches required in an FPGA is measured.

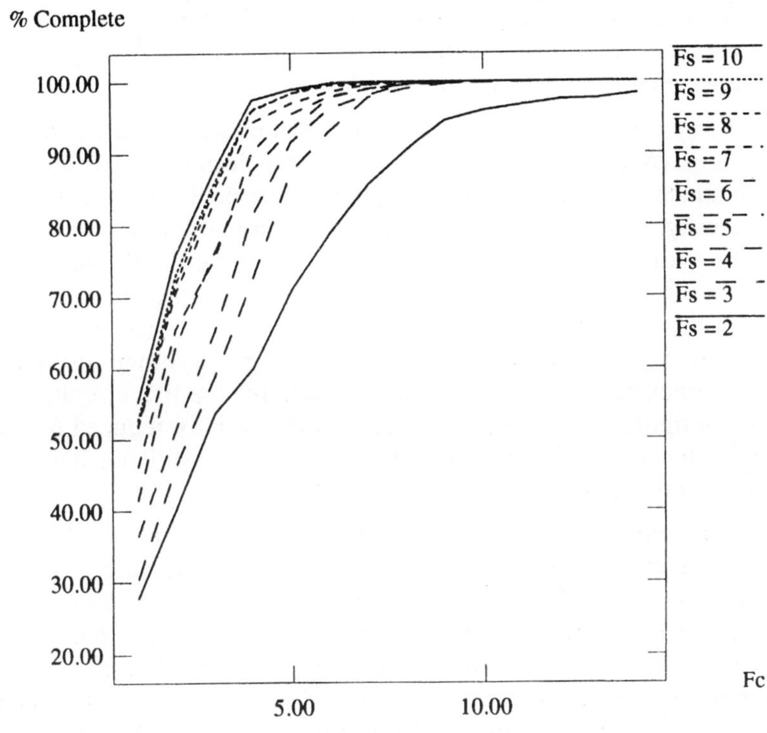

Figure 6.9 - *Percent Routing Completion vs. Fc, Circuit BNRE.*

6.4.1 Effect of Connection Block Flexibility on Routability

Figure 6.9 is a plot of the percentage of successfully routed connections versus connection block flexibility, F_c, for the circuit BNRE. Each curve in the figure corresponds to a different value of switch block flexibility, F_s. The lowest curve represents the case $F_s = 2$ and the highest curve to $F_s = 10$. The number of tracks, W, is set to 14, which is two greater than W_g, the minimum possible number of tracks as indicated by the global router. The value of $W = W_g + 2$ was chosen to give the detailed router a reasonable chance of success. Using a higher or lower value of W would shift the curves slightly upward or downward, respectively. Figure 6.9 indicates that the routing completion rate is very low for small values of F_c and only achieves 100% when F_c is at least one-half of W. The figure also shows that increasing the switch block flexibility improves the completion rate at a given F_c, but to get near 100% the value of F_c must always be high (above 7 for this circuit).

Table 6.2 summarizes the results for the other circuits. It gives the minimum values of F_c and F_c/W required to achieve 100% routing completion for each circuit, for nine values of F_s. W is fixed at $W_g + 2$, in all cases, to give a reasonable chance for success.

The key observation from the data of Table 6.2 is that there appear to be minimum values of F_c and F_c/W below which circuits are not routable. However, since this data is based on a fixed value of $W = W_g + 2$, it is interesting to investigate whether F_c or F_c/W can be reduced if W is not fixed. To study this, a similar experiment was conducted in which W was allowed to vary to a maximum of $3 \times W_g$. Again, the experiments measure the minimum possible values of F_c and F_c/W for which 100 percent routing can be achieved, for a range of values of F_s. The results are shown in Figure 6.10, which for conciseness gives the average results for the five circuits. The left curve in the figure shows that F_c/W can be substantially reduced by allowing W to vary, but the curve to the right shows that F_c still reaches about the same minimum value.

To see why there exists a minimum value of F_c below which circuits are not routable, consider the following discussion concerning C block topology. Assume that a C block must connect n logic block pins to a set of tracks, and that some pin, p_i, must be able to connect to all of pins p_j, $1 \le j \le n$. Some connections between these pin pairs will occur within one C block and others will involve two different C blocks. To simplify the analysis, assume that $F_s \le 3$, so that no "jogging" is allowed among the tracks. As the discussion in Section 6.1.2.1 showed, the C block topology must provide at least one common track that connects to both p_i and each p_j. To accomplish this, the design of the topology may:

(1) Attach two switches to each of n different tracks, such that each track connects one p_j to p_i. In terms of Section 6.1.2.1, this corresponds to spreading the switches evenly across the tracks.

(2) Attach n switches to any one track, such that p_i can connect to any p_j on that track.

Circuit	W	Fs	100% Fc	Fc/W
BUSC	11	3	9	0.82
BUSC	11	4	7	0.64
BUSC	11	5	7	0.64
BUSC	11	6	6	0.54
BUSC	11	7	6	0.54
BUSC	11	8	5	0.45
BUSC	11	9	6	0.54
DMA	12	3	8	0.67
DMA	12	4	7	0.58
DMA	12	5	7	0.58
DMA	12	6	7	0.58
DMA	12	7	7	0.58
DMA	12	8	5	0.42
DMA	12	9	5	0.42
BNRE	14	3	12	0.86
BNRE	14	4	11	0.79
BNRE	14	5	10	0.71
BNRE	14	6	9	0.64
BNRE	14	7	10	0.71
BNRE	14	8	8	0.57
BNRE	14	9	8	0.57
DFSM	13	3	9	0.69
DFSM	13	4	9	0.69
DFSM	13	5	9	0.69
DFSM	13	6	8	0.62
DFSM	13	7	8	0.62
DFSM	13	8	7	0.54
DFSM	13	9	7	0.54
Z03	13	3	10	0.77
Z03	13	4	9	0.69
Z03	13	5	9	0.69
Z03	13	6	9	0.69
Z03	13	7	7	0.54
Z03	13	8	7	0.54
Z03	13	9	7	0.54

Table 6.2 - *Minimum F_c Required for 100% Completion.*

(3) Use a combination of options (1) and (2).

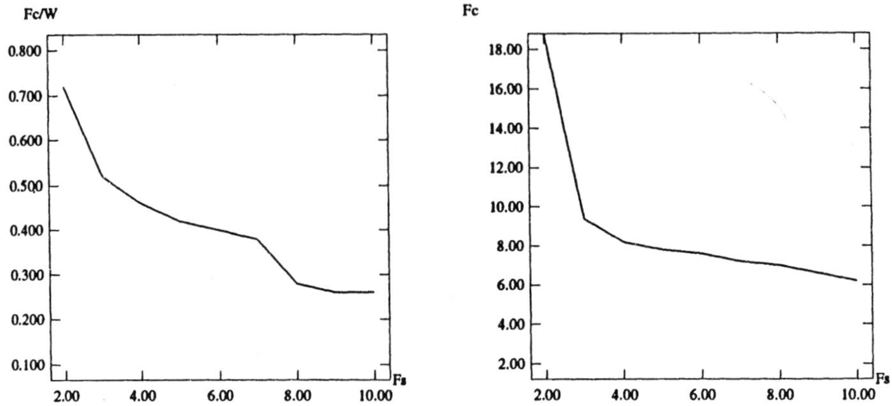

Figure 6.10 - $\frac{F_c}{W}$ *vs.* F_s *and* F_c *vs.* F_s*, Variable W.*

Option (1) leads directly to a minimum value for F_c because it entails attaching n switches to p_i. The effect of option (2) is more subtle, as discussed below. Consider Figure 6.11, which shows a C block topology in which each pin connects to exactly the same F_c tracks as every other pin. The figure shows that, with this topology, when one pin is connected to a track, one choice of track is eliminated for every other pin. In this scenario, it follows that the minimum possible value of F_c is determined by the maximum number of pins that are connected at any C block.

A more realistic C block, such as the one that was shown in Figure 6.6, is based on option (3). This means that a combination of the effects of options (1) and (2) determines a minimum value for F_c. The key to this discussion is that any realistic C block must provide connections between a number of different pairs of pins and this leads directly to a minimum possible value for F_c.

Note that the minimum value of F_c can be reduced slightly by increasing F_s to be above three, because this increases the connectivity between pairs of pins by allowing jogging from one track to another. However, this only affects connections that involve two different C blocks, and since some connection's pins are both within one C block, an absolute minimum value exists for F_c.

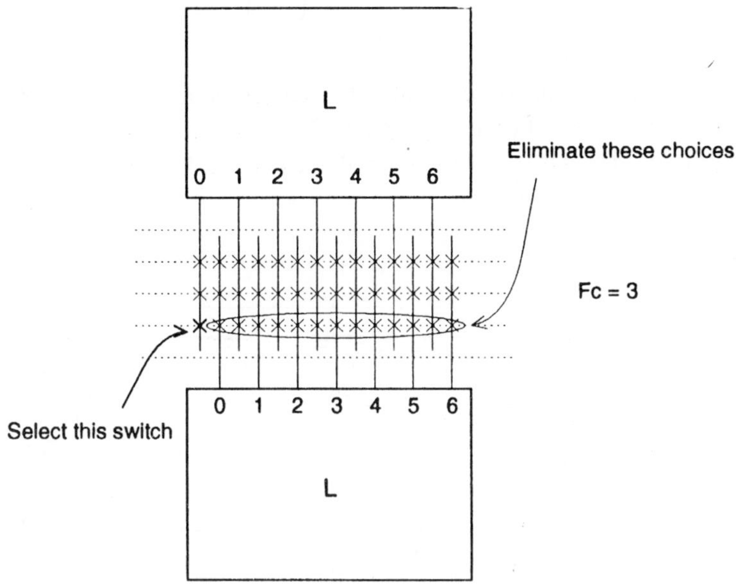

Figure 6.11 - *Connecting One Pin Eliminates One Choice for Every Other.*

6.4.2 Effect of Switch Block Flexibility on Routability

Figure 6.12 is a plot of the percentage routing completion versus switch block flexibility, F_s. Each curve in the figure corresponds to a different value of F_c, with the lowest curve representing $F_c = 1$ and the highest curve corresponding to $F_c = W$. This plot is for the circuit BNRE, with W set to 14. The figure shows that if F_c is high enough, then very low values of F_s can achieve 100% routability. These F_s values are low in comparison with the maximum possible value of F_s, which is $3 \times W$. For the results in Figure 6.12 this maximum is 42, whereas 100% routing completion is often achieved for $F_s = 3$. This makes intuitive sense because even for $F_s = 3$ every track that passes through a particular C block is guaranteed to connect to one other track at every other C block. To further quantify the effect of F_s on routability, consider the connection of two logic block pins that are separated by n S blocks. The number of tracks connectable at the first logic block pin is F_c and the number of paths available to reach the connection block adjacent to the second logic block is

$$\# Paths = F_c \times \left(\frac{F_s}{3}\right)^n$$

Using the average value of n of about 3 for typical circuits, if $F_s = 3$ and $F_c = 10$, then there are 10 paths available. If F_s is increased to 6, there are 80 paths available. Thus a small increase in F_s greatly increases the number of paths, and hence the routability.

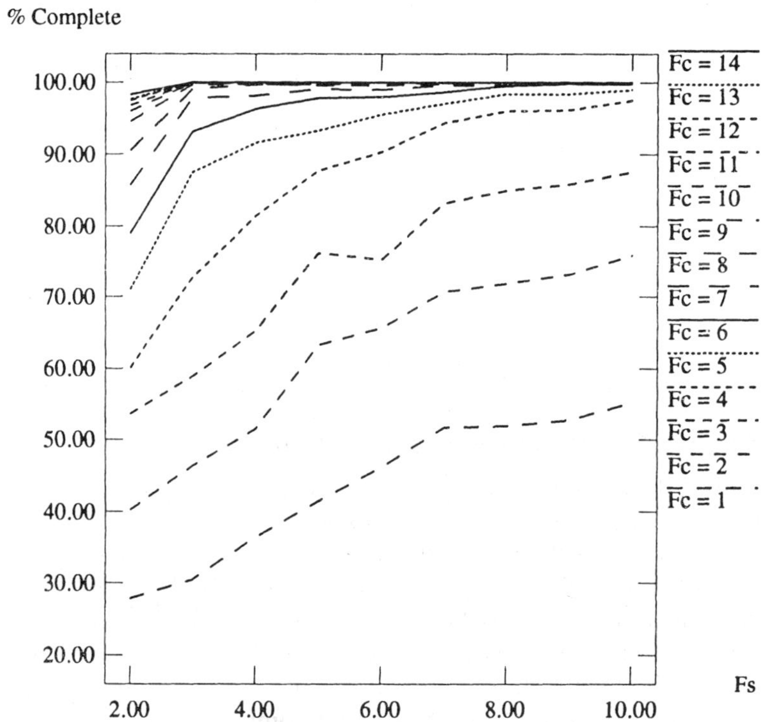

Figure 6.12 - *Percent Routing Completion vs. Fs, Circuit BNRE.*

6.4.3 Tradeoffs in the Flexibilities of the S and C Blocks

Figures 6.9 and 6.12 can be combined in three-dimensions to show that a tradeoff exists between the flexibilities of the S blocks and the C blocks. This is illustrated by the three-dimensional surface plot in Figure 6.13. The plot shows, for example, that a decrease along the F_c axis can be compensated for by an increase along the F_s axis, and vice-versa. This can also be

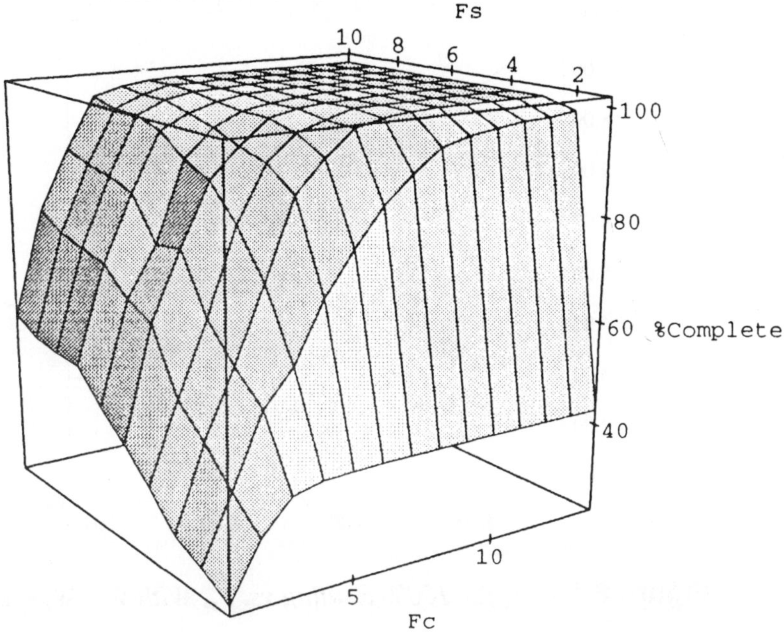

Figure 6.13 - *C Block and S Block Tradeoff.*

seen in Figure 6.14, which is a plot of the minimum value of F_c (averaged over the five circuits) for which 100 percent routing can be attained for a range of values of F_s. In this plot, W is constant for each circuit, at two higher than W_g. Note that there is no data point for the case $F_s = 2$ in the figure because the circuits are not routable for that S block flexibility with $W = W_g + 2$. The slope of the curve in Figure 6.14 will flatten for higher values of F_s since there exists a minimum value of F_c for each circuit, as was discussed in Section 6.4.1.

Note that Figure 6.14 alone does not provide enough information to choose the best values of F_c and F_s because it is based on the fixed value of W. The required value of W for different values of F_c and F_s is the subject of the following section.

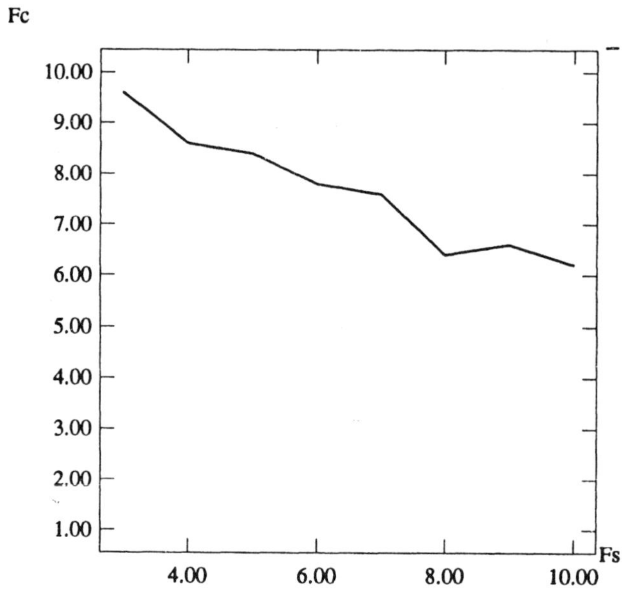

Figure 6.14 - *F_c for 100% Routing vs. F_s, With $W = W_g + 2$.*

6.4.4 Track Count Requirements

This section investigates the effect of the S and C block flexibilities on routability by measuring the number of tracks per channel, W, required to route 100 percent of a circuit's connections for specific values of F_s and F_c. For these experiments, the detailed router was invoked repeatedly over a range of values of W, from W_g (the maximum channel density of the circuit) to a maximum of $3 \times W_g$, until 100 percent routing was achieved. Each routing architecture is assessed by comparing how close the required W is to W_g. For these experiments the flexibility of the C block is expressed as the ratio of F_c / W - as W is changed, the ratio is kept constant. This provides a convenient way to coordinate changes in W with the number of routing switches in a C block. This is automatic for an S block, since F_s applies to each track.

Table 6.3 shows the number of tracks required to achieve 100 percent routing completion for circuit BNRE over a set of FPGA routing architectures. Each entry in the table corresponds to a different pair of $(F_s, F_c/W)$ values. The value "nr" in the table means that 100 percent routing was not achieved. Since W_g for this circuit is 12, the table shows that even with low flexibilities it is possible to achieve 100 percent routing using very near to the minimum possible number of tracks.

Table 6.4 summarizes the results for all the circuits. Each entry in this table is the average over all the circuits of the number of tracks that are required *in excess* of the minimum (W_g) to route 100 percent of the connections. These results show that with very low flexibilities it is possible to achieve a number of tracks only slightly greater than the minimum. In particular, for $F_s \geq 3$ and $(F_c/W) \geq 0.6$, excluding the case ($F_s = 3$, $(F_c/W) = 0.6$), the number of tracks required in excess of the minimum is no greater than three.

	Fc/W							
Fs	0.3	0.4	0.5	0.6	0.7	0.8	0.9	1.0
2	nr	nr	nr	nr	nr	23	23	17
3	nr	nr	nr	18	14	13	13	13
4	nr	nr	18	14	13	13	13	13
5	nr	nr	18	14	13	12	12	12
6	nr	21	16	14	14	12	13	13
7	nr	19	17	14	12	12	12	12
8	nr	19	15	13	12	12	12	12
9	23	17	13	13	12	12	12	12
10	19	16	13	13	12	12	12	12

Table 6.3 - *Track Count Requirements for BNRE (Minimum = 12).*

6.4.5 Architectural Choices

The choice of a particular FPGA routing architecture must be based on the cost of its implementation in terms of area and delay. Although these depend on the technology used for the routing switches, it is possible to make general comments because, regardless of their implementation, routing switches consume area and cause time delays. It is instructive to examine the total number of routing switches required by different routing architectures. The number of switches in a C block and an S block depends on the number of logic block pins (P), the number of sides that each pin appears on (T), and on the parameters F_c, F_s, and W, according to the following equations:

$$\#\,Switches\ in\ Connection\ Block = \frac{1}{2} \times T \times P \times F_c$$

$$\#\,Switches\ in\ Switch\ Block = 2 \times F_s \times W$$

For the results presented here *P* is 7, *T* is 2, and the last three parameters are read from the equivalent of Table 6.3 for each circuit. As the flexibility of the routing structures increases (F_s and F_c) the number of switches per track increases, but the number of tracks per channel may decrease, as shown in Table 6.3. Hence there should be an architecture that exhibits a minimum total number of switches. Table 6.5 gives the number of switches *per tile* required for circuit BNRE for each FPGA routing architecture. A tile is the section of the FPGA that would be replicated across the entire chip, and includes the logic block, two connection blocks and one switch block.

	Fc/W							
Fs	0.3	0.4	0.5	0.6	0.7	0.8	0.9	1.0
2	nr	nr	nr	nr	nr	11.2	10.8	9.0
3	nr	nr	nr	4.6	2.4	1.2	0.8	0.8
4	nr	nr	6.2	3.0	1.6	0.6	0.6	0.6
5	nr	nr	4.2	2.4	1.0	0.4	0.2	0.2
6	nr	7.6	3.8	1.8	0.8	0.4	0.4	0.4
7	nr	5.2	3.4	1.4	0.2	0.2	0.2	0.2
8	nr	4.4	1.4	0.6	0.2	0.0	0.4	0.2
9	10.0	4.2	1.0	0.4	0.2	0.2	0.0	0.0
10	8.0	3.2	1.4	0.6	0.2	0.0	0.0	0.0

Table 6.4 - *Average Excess Track Count Requirements over all Circuits.*

As Table 6.5 shows, flexibilities of $F_s = 3$ and $(F_c/W) = 0.7$ achieve a minimum number of switches for this circuit, at 221. Note that several neighboring architectures have similar switch counts. For all of the test circuits the minimum number of switches was between 172 and 223, and occurred when the architecture's parameters were in the range $3 \le F_s \le 4$ and $0.7 \le (F_c/W) \le 0.8$.

6.5 Conclusions

This chapter has explored the relationships between the flexibility of routing architectures and routability in FPGAs. The principal conclusions

	Fc/W							
Fs	0.3	0.4	0.5	0.6	0.7	0.8	0.9	1.0
2	nr	nr	nr	nr	nr	349	381	306
3	nr	nr	nr	259	221*	223	241	260
4	nr	nr	270	229	231	249	267	286
5	nr	nr	306	257	257	254	271	288
6	nr	369	304	285	305	278	319	338
7	nr	372	357	313	285	302	319	336
8	nr	410	345	317	309	326	343	360
9	510	401	325	343	333	350	367	384
10	459	409	351	369	357	374	391	408

Table 6.5 - *Average Number of Switches per Tile for Each Architecture.*

are that connection blocks should have high flexibility to achieve high percentage routing completion, but that a relatively low flexibility is sufficient in the switch blocks. Furthermore, with low flexibilities the number of tracks per channel required is very close to the minimum. Finally, it has been shown that routing architectures with these properties yield the lowest total number of routing switches.

This architectural study has been performed using an experimental approach, in which CAD tools are used to implement circuits and the effects of flexibility on routability are measured by the outcomes of the experiments. FPGA routing architectures are studied differently in the next chapter, in which a stochastic model is used in a theoretical study of flexibility and routability.

CHAPTER 7
A Theoretical Model for FPGA Routing

The previous chapter presented an experimental study of the effect of the flexibility of the routing architecture in a symmetrical FPGA on the routability of circuits. Flexibility was defined to be a function of the total number of routing switches and wire segments and routability was measured as the percentage of a circuit's connections that could be successfully routed by CAD tools. This chapter presents a similar study of symmetrical FPGAs, but uses theory rather than experiments. The approach taken is to develop a mathematical model based on probability theory to represent routing in FPGAs as a random process [Brow92]. The model uses simple parameters for the FPGA itself and a circuit to be routed, and allows a range of routing architectures to be specified. Two purposes are served by this study: it confirms the experimental results that were presented in Chapter 6, and it provides a basis for future studies of FPGA routing architectures. The latter goal is the most significant one since a theoretical study is often much easier to perform than one that involves time-consuming experiments and possibly the need to develop new CAD programs.

The stochastic model that will be described assumes that a two-stage routing approach of global routing followed by detailed routing is used (see Section 5.2). The global routing stage is represented in the model by making assumptions, which are described in Section 7.2, concerning the solution that a global router would produce. Assuming that a connection is assigned a single global route, the stochastic model calculates the probability that the connection could be successfully routed, given the size of the FPGA and its routing architecture, and considering the number of other connections that have been previously routed. As Section 7.2 describes, once all of the connections in a circuit have been processed in this way, routability can be calculated by the average probability of routing success.

7.1 Architectural Assumptions for the FPGA

FPGAs are characterized in the stochastic model using the same basic assumptions that appear in Chapters 5 and 6. For ease of reference, some information discussed in those chapters is repeated here. An example FPGA is illustrated in Figure 7.1, which shows a square array of logic blocks, with *N* blocks per side. Note that the grid coordinates in Figure 7.1 number only the logic blocks, which differs from figures shown in earlier chapters. This change has been made because this labelling scheme is more convenient for the stochastic model.

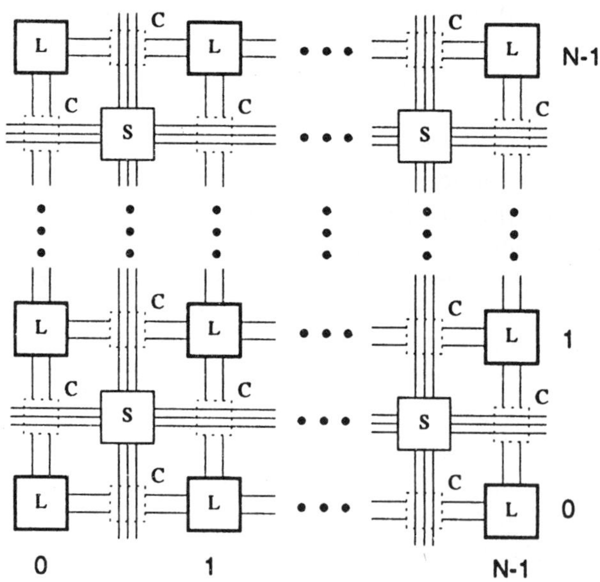

Figure 7.1 - *An N×N FPGA.*

No assumptions are necessary about the internal details of the logic blocks, except that each block has some number of pins that are connected to the routing channels by routing switches. Routing switches are contained within C blocks and S blocks. It is assumed that tracks are hard-wired straight through the C blocks, without passing through routing switches, but that a routing switch is always needed to pass through an S block. This means that all tracks are composed of wire segments that span the length of one logic block. The switches in the C blocks are used to connect the logic block pins to the wire segments, and the S block switches provide connections from one wire segment to another. The flexibility of a C block is defined by F_c, which specifies the number of tracks that each logic block pin can connect to. The flexibility of an S block is given by F_s, which defines the number of other wire segments that a wire segment entering an S block can be connected to.

7.2 Overview of the Stochastic Model

In the stochastic model, it is assumed that a circuit with a total of C_T two-point connections is to be routed in an FPGA with $N \times N$ logic blocks. The length of each connection is drawn from a probability distribution, P_L. It will later be necessary to choose a specific distribution for P_L. In Section 7.4.4, it is assumed that P_L is geometric, with mean length $\bar{R}$. This assumption is taken from earlier work on the stochastic modelling of two-dimensional arrays of connected cells [Heller84] [ElGa81], and has the following physical interpretation in an FPGA: at each C block along the path of a connection, the connection will terminate (at a logic block) with probability $\frac{1}{\bar{R}}$ and will continue (to the next C block) with probability $1 - \frac{1}{\bar{R}}$.

The C_T connections are individually referred to as $C_1, C_2, \ldots, C_{C_T}$ and the statistical event that each connection is successfully routed is called $R_{C_1}, R_{C_2}, \ldots, R_{C_{C_T}}$. The key to the stochastic model is the calculation of the probabilities of $R_{C_1}, R_{C_2}, \ldots, R_{C_{C_T}}$. Recall that routability is defined as the percentage of the connections in a circuit that can be successfully routed. In terms of $R_{C_1}, R_{C_2}, \ldots, R_{C_{C_T}}$ this corresponds to the ratio of the expected number of successfully routed connections to the total number of connections, C_T. Thus, routability is the average probability of completing a connection and can be calculated in the stochastic model according to

$$Routability = \frac{1}{C_T} \sum_{i=1}^{C_T} P(R_{C_i}),$$

where $P(R_{C_i})$ is the probability of successfully routing C_i.

7.2.1 Model of Global Routing and Detailed Routing

In order to use a key research result by El Gamal [ElGa81] to predict the densities of the routing channels in an FPGA, the following assumption is made concerning the way in which a global routing algorithm would assign the connections in a circuit to the routing channels. It is assumed that each connection is assigned a single path through the routing channels in such a way that the number of connections per routing channel is Poisson distributed. Section 7.3 justifies this assumption and illustrates its use.

In the stochastic model, the detailed routing of an FPGA is represented as a random process. Based on the assumption that a connection is assigned a single path through the routing channels, the probability of successfully performing the detailed routing of the connection is calculated, accounting for the number of tracks per routing channel, the flexibilities of the C and S blocks, and the side-effects that the routing of one connection has on others.

In Chapter 5, it was shown that a key issue in the detailed routing of symmetrical FPGAs is how the routing of one connection may affect other connections. To compute the value of each $P(R_{C_i})$, it is necessary for the stochastic model to account for these effects. To accomplish this, the model 'routes' the connections in a serial manner and predicts the effect that each successfully routed connection has on the densities of the routing channels. By this mechanism, the probability of routing each subsequent connection is influenced because there are more connections in a channel to compete with. The next section shows how El Gamal's results can be used to calculate channel densities and following this, the probability formulas for $P(R_{C_i})$ are derived.

7.3 Previous Research for Predicting Channel Densities

In [ElGa81], a stochastic model is developed to predict the wiring requirements of Master Slice Integrated Circuits that have a two-dimensional array of identical cells, with horizontal and vertical routing channels between the rows and columns of cells. The model divides the channels into segments that span the length or width of one cell and it is assumed that all interconnections start at one cell and travel a minimum distance through the channel segments to another cell. It is further assumed that the number of connections per cell can be drawn independently from a Poisson distribution, with parameter λ, where λ is defined as the ratio of the total number of connections in a circuit to the total number of cells in the array. The average connection length, in number of cells traversed, is called $\bar{R}$. The paper also makes assumptions about the trajectories of connections, but these assumptions are not necessary for the results that are quoted here.

[ElGa81] shows that under the above assumptions, in an array that has $N \times N$ routing channels, the densities of the channel segments will be Poisson distributed, with the average density given by $\frac{\lambda \bar{R}}{2}$. This result holds as long as $\bar{R} < \infty$, independent of N, and provides a convenient method of predicting channel densities.

7.3.1 Predicting Channel Densities in FPGAs

Although the results in [ElGa81] were developed for Master Slice circuits, they can also be applied to the FPGAs considered here, because both types of devices are based on a two-dimensional array of identical cells. The definitions of the routing channels differ, but these differences can be ignored since the tracks consist of short segments that span only one logic block in both cases.

Having made these assumptions, it is convenient to predict channel densities in FPGAs using El Gamal's result. The accuracy of the predictions can be checked by comparing the ideal Poisson distribution with mean $\frac{\lambda \bar{R}}{2}$ to the distribution of channel densities in real FPGA circuits. Figure 7.2 shows the result of such a comparison for one of the circuits (called BNRE) that was used in the experiments presented in Chapter 6. It compares the channel densities measured from the circuit to the ideal Poisson distribution, with the values of λ and $\bar{R}$ measured from the circuit. As the figure shows, the actual channel densities are surprisingly close to the Poisson predictions. Comparisons with the other circuits used in Chapter 6 show similar results.

It is interesting to discuss a physical interpretation of the Poisson distribution in this context. Assume that an FPGA has W tracks in each routing channel and consider a specific channel segment. For each of the W tracks, define p_i as the probability of the statistical event that the track would be occupied if a circuit were routed in the FPGA. If $W = 1$, there will be a probability, p_1, that the track will be occupied by some connection. If $W=2$, then there will be a probability, p_2, that each of the two tracks will be occupied by some connection and $p_2 < p_1$. Extending this to the general case, if $W=n$, then each track will be occupied with probability p_n, and $p_n < p_{n-1} < \cdots < p_2 < p_1$. Furthermore, as $n \to \infty$, $p_n \to 0$. Since p_n is small in the limiting case, the event that a track is used is a rare event and the number of these events (density) can be approximated by the Poisson distribution.

In FPGAs in which the tracks consist of segments that span multiple logic blocks, El Gamal's result is probably not an accurate approximation of channel densities. In such cases, a different method of calculating densities would be needed. However, in this chapter it is assumed that all FPGAs being considered have tracks consisting of only short segments.

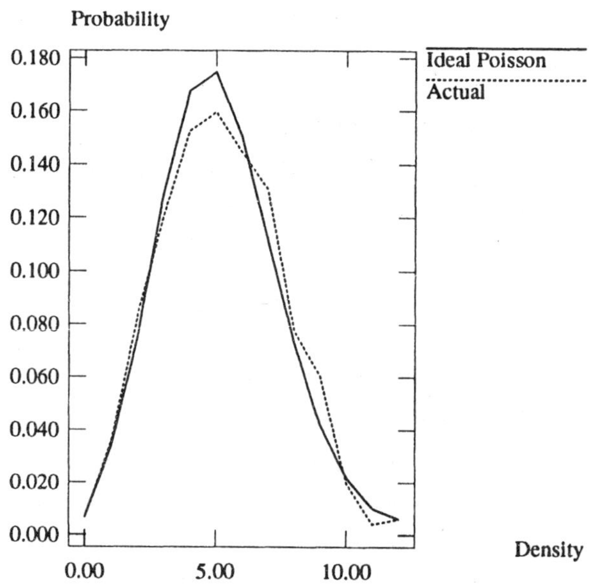

Figure 7.2 - *Predicted versus Actual Channel Densities.*

7.4 The Probability of Successfully Routing a Connection

This section derives analytic expressions for calculating the probability of successfully performing the detailed routing of a single connection in an FPGA. As an example of a connection, consider Figure 7.3. The figure shows a connection, C_i, that starts at logic block (x_1,y_1) and travels through routing channels to logic block (x_2,y_2). The length of C_i is defined in terms of logic block hops as $LC_i = |x_1 - x_2| + |y_1 - y_2|$ (to be consistent with [ElGa81]). Also, the number of S blocks that C_i passes through is given by $LC_i - 1$. To define the probability, $P(R_{C_i})$, of successfully routing C_i, assume that $LC_i = n+1$. The statistical event that corresponds to this assumption is written as L_{n+1}. Also define the following events:

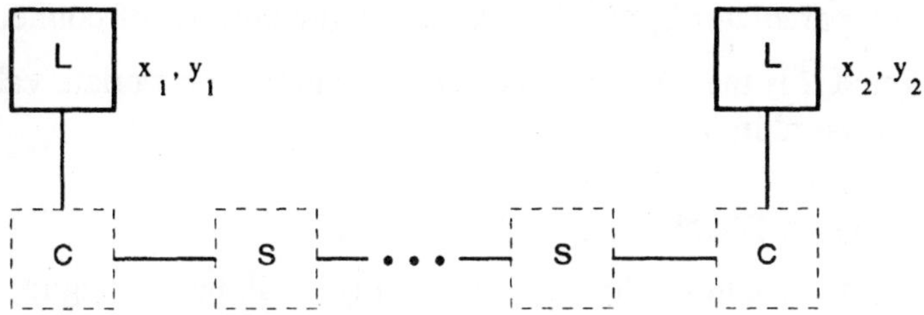

Figure 7.3 - *A Typical Connection.*

- X_1 - the event that the logic block pin associated with C_i at (x_1,y_1) can connect to at least one track at the first C block. Note that there are, by definition, F_c tracks that can connect to the logic block pin, but any number of them may already be used by other connections that have been previously 'routed'.
- $S_1, S_2, \cdots, S_n$ - the events that C_i can successfully reach at least one track on the outgoing side of the first, second, up to the nth S block. There are $LC_i - 1$ such events for C_i.
- X_2 - the event that at least one of the tracks that are available to C_i at the last C block can be connected to the appropriate logic block pin at (x_2,y_2).
- R_{C_i} - the event that C_i can be successfully routed.

Since C_i is successfully routed only if all of the events $X_1, S_1, S_2, ..., S_n, X_2$ occur, then

$$R_{C_i} \mid L_{n+1} = X_1 \cap S_1 \cap S_2 \cdots \cap S_n \cap X_2$$

and the probability of successfully routing C_i is given by

$$P(R_{C_i} \mid L_{n+1}) = P(X_1 \cap S_1 \cap S_2 \cdots \cap S_n \cap X_2)$$

$$= P(X_1)P(S_1 \mid X_1)P(S_2 \mid S_1 \cap X_1) \cdots P(S_n \mid S_{n-1} \cap \cdots \cap S_1 \cap X_1)$$

$$\cdot P(X_2 \mid S_n \cap \cdots S_1 \cap X_1). \qquad 7.1$$

Since the events $X_1, S_1, S_2, ..., S_n$, and X_2 are not independent, it is necessary to find formulas for each of the terms in Equation 7.1. This is accomplished in the following sections by developing expressions that account for the flexibilities of the C and S blocks (F_c and F_s), the number of tracks per routing channel (W), and the densities of the routing channels. As discussed in Section 7.3, channel density is approximated by the Poisson distribution with parameter $\lambda_g = \frac{\lambda \bar{R}}{2}$, where λ is the number of connections per logic block and $\bar{R}$ is the average connection length. Appropriate values for λ and $\bar{R}$ are discussed in Section 7.5.

7.4.1 The Logic Block to C Block Event

The event X_1 can be depicted pictorially as shown in Figure 7.4. The figure shows a routing channel with W tracks and a logic block pin that can connect to F_c of the tracks, via routing switches (shown by an X). The figure also shows a set of D tracks, drawn as dashed lines, that are already occupied by previously routed connections. In the figure, $W=10$, $F_c=5$, and $D=5$. The event X_1 can then be viewed as a random process in which a logic block pin can connect to any of the unused tracks where there are switches.

To derive a formula for $P(X_1)$, it is convenient to define the event NONE as the opposite of X_1 - i.e. $P(X_1)=1-P(NONE)$. $P(NONE)$ can be calculated by observing that the value of D is Poisson distributed and that a Poisson distribution is infinitely divisible. This means that rather than considering a Poisson process over W tracks, with mean λ_g, it is sufficient to deal with a smaller Poisson process over F_c tracks, with mean $\lambda_g \frac{F_c}{W}$. The probability $P(NONE)$ is then easily calculated as the probability that $D = F_c$ within

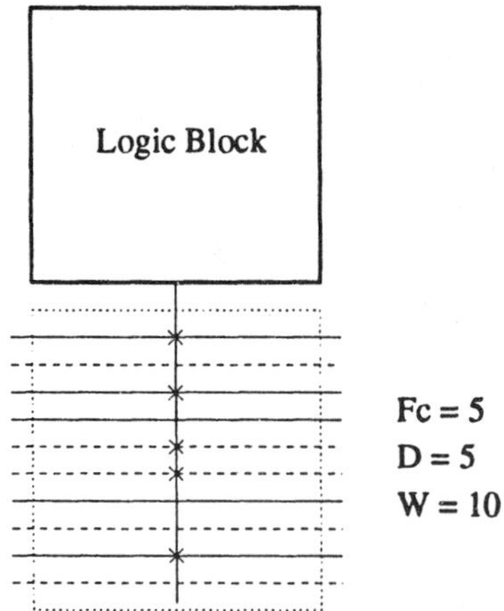

Figure 7.4 - *The Event X_1.*

the smaller Poisson process, or

$$P(NONE) = p(\lambda_g \frac{F_c}{W}, F_c). \tag{7.2}$$

$P(X_1)$ is then given by

$$P(X_1) = 1 - P(NONE) = 1 - p(\lambda_g \frac{F_c}{W}, F_c). \tag{7.3}$$

Note that Equation 7.3 is an approximation because the Poisson distribution has an infinite tail, but the equation will only be evaluated for values of F_c as large as W. The infinite tail means that there is a non-zero probability of channel densities above W but for practical values of W this error is very small and can be ignored.

Equation 7.3 depends upon the channel densities being Poisson distributed so that the property of being infinitely divisible can be used. This means that the resulting expression is only applicable for FPGAs in which the channel densities can be approximated by the Poisson distribution. For FPGAs that do not satisfy this assumption, a different approach to solving for $P(X_1)$ must be followed. This same statement also applies for other expressions that are developed later in this chapter. Examples of FPGAs that do not follow the Poisson assumption of channel densities are those that have tracks with segments that span multiple logic blocks. Some of the required modifications for such cases can be found in [Brow92].

Equation 7.3 calculates $P(X_1)$ based on the relationship between the event X_1 and the event *NONE*. An alternative is to calculate $P(X_1)$ directly by defining $A_1^{X_1}, A_2^{X_1}, ..., A_{F_c}^{X_1}$ as the events that X_1 occurs with exactly $1, 2, ..., F_c$ available tracks. Using this approach,

$$X_1 = A_1^{X_1} \cup A_2^{X_1} \cup \cdots \cup A_{F_c}^{X_1}$$

and since $A_1^{X_1}, A_2^{X_1}, ..., A_{F_c}^{X_1}$ are mutually exclusive,

$$P(X_1) = P(A_1^{X_1}) + P(A_2^{X_1}) + \cdots + P(A_{F_c}^{X_1}).$$

Although $P(X_1)$ can be calculated using Equation 7.3, each of $P(A_a^{X_1})$ will be required in the next section, so they are derived here. Consider the general case of X_1 occurring with exactly (a) available tracks, and the corresponding event $A_a^{X_1}$. Since only the F_c tracks that have switches are of interest, then if exactly (a) tracks are available, $F_c - a$ tracks must already be occupied, meaning that $D = F_c - a$. Following the approach used for Equation 7.2, the probability that $D = F_c - a$ can be calculated directly from a scaled-down Poisson process for the F_c tracks, which gives

$$P(A_a^{X_1}) = p(\lambda_g \frac{F_c}{W}, F_c - a). \qquad 7.4$$

As a check, it is easily verified that $P(NONE)$ can be obtained using equation 7.4 by setting $a=0$, which must be true since $P(NONE) \equiv P(A_0^{X_1})$. Finally,

$$P(X_1) = \sum_{a=1}^{F_c} P(A_a^{X_1}) = \sum_{a=1}^{F_c} p(\lambda_g \frac{F_c}{W}, F_c - a).$$

7.4.2 The S Block Events

All of the events that are associated with S blocks can be treated in a uniform way. This section first derives probability formulas for $S_1 \mid X_1$ and then shows how the resulting expressions can be applied to subsequent S blocks.

7.4.2.1 The First S Block Event, for $F_s = 3$

Since $P(S_1 \mid X_1)$ will be affected by the flexibility of the S block, it is convenient to assume a specific value of F_s. In the following derivation, the case $F_s = 3$ is assumed. This is the easiest case to handle because it means that each wire segment that enters an S block can connect to exactly one wire segment on each other side. Also, the derivation need not be concerned with whether a connection turns or passes straight through an S block since the effect is the same in both cases.

The event $S_1 \mid X_1$ is depicted in Figure 7.5, which shows an S block and a routing channel that has W tracks. The figure shows a set of A^{X_1} tracks, drawn as bold lines, that are available at the incoming side of the S block and a set of D tracks, drawn as dashed lines on the outgoing side of the S block, that are already used by other connections. In the figure, $D=4$, $W=10$, and $A^{X_1}=3$. Note that setting A^{X_1} to three corresponds to the event $A_3^{X_1}$, from Section 7.4.1. Figure 7.5 uses dotted lines to indicate S block switches and shows that each track on the incoming side of the S block can be connected to one other track on the outgoing side. The event can then be considered to be a random process in which each of the A^{X_1} incoming tracks can connect to one track on the outgoing side of the S block, as long as that outgoing track is not among the D used tracks. In other words, given that there are A^{X_1} tracks that are available on the incoming side of the S block, it is necessary to find the probability that one or more of these tracks are also available on the outgoing side.

The event $S_1 \mid X_1$ can occur with one or more available outgoing tracks. To calculate $P(S_1 \mid X_1)$, define $A_1^{S_1}, \ldots, A_{F_s}^{S_1}$ as the events that $S_1 \mid X_1$ occurs with exactly 1, 2, ..., F_c available tracks on the outgoing side. Since

$$S_1 \mid X_1 = A_1^{S_1} \cup \cdots \cup A_{F_c}^{S_1}$$

and $A_1^{S_1}, \ldots, A_{F_c}^{S_1}$ are mutually exclusive

$$P(S_1 \mid X_1) = P(A_1^{S_1}) + \cdots + P(A_{F_c}^{S_1}) . \qquad 7.5$$

To solve for each term in this summation, consider the general case where $S_1 \mid X_1$ occurs with exactly k available outgoing tracks. The corresponding event is written $A_k^{S_1}$. The probability of $A_k^{S_1}$ will depend on the number of tracks available on the incoming side, given by A^{X_1}, and on the value of D. Assume a specific value of $A^{X_1} = a$. Since X_1 is known to have occurred, this corresponds to assuming that X_1 occurred with exactly (a) available tracks. The appropriate statistical event for this assumption is then written as $A_a^{X_1} \mid X_1$. If exactly (k) tracks are available, this implies that $k-a$ tracks are already used, meaning that $D = k - a$. The probability $P(A_k^{S_1} \mid (A_a^{X_1} \mid X_1))$ can be calculated by again observing that D is Poisson distributed, which means that the distribution is divisible. In this case, the only tracks that are of interest on the outgoing side of the S block are the set of (a) tracks that can be reached from the (a) tracks available on the incoming side. Thus, the scaled-down Poisson process that should be used to calculate D has a mean given by $\lambda_g \frac{a}{W}$, so that

$$P(A_k^{S_1} \mid (A_a^{X_1} \mid X_1)) = p(\lambda_g \frac{a}{W}, a - k) . \qquad 7.6$$

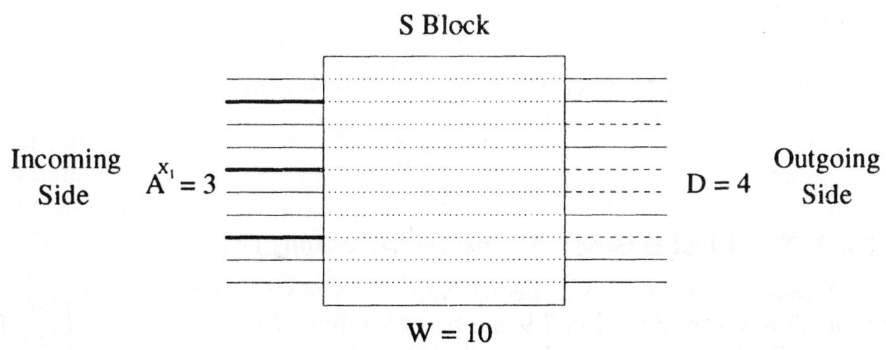

Figure 7.5 - *The Event S_1.*

Next, consider the events $(A_1^{X_1} \mid X_1), \ldots, (A_{F_c}^{X_1} \mid X_1)$ corresponding to the possible values of A^{X_1}. Since the occurrence of $A_k^{S_1}$ implies exactly one of $(A_1^{X_1} \mid X_1), \ldots, (A_{F_c}^{X_1} \mid X_1)$, then

$$A_k^{S_1} = (A_k^{S_1} \cap (A_1^{X_1} \mid X_1)) \cup \cdots \cup (A_k^{S_1} \cap (A_{F_c}^{X_1} \mid X_1))$$

and since $(A_1^{X_1} \mid X_1), \ldots, (A_{F_c}^{X_1} \mid X_1)$ are mutually exclusive

$$P(A_k^{S_1}) = P(A_k^{S_1} \cap (A_1^{X_1} \mid X_1)) + \cdots + P(A_k^{S_1} \cap (A_{F_c}^{X_1} \mid X_1)).$$

Using the relation $P(X \cap Y) = P(X)P(Y \mid X)$, get

$$P(A_k^{S_1}) = P(A_1^{X_1} \mid X_1)P(A_k^{S_1} \mid (A_1^{X_1} \mid X_1)) + \cdots$$

$$+ P(A_{F_c}^{X_1} \mid X_1)P(A_k^{S_1} \mid (A_{F_c}^{X_1} \mid X_1)).$$

The terms $P(A_k^{S_1} \mid (A_a^{X_1} \mid X_1))$ are given by equation 7.6, so that

$$P(A_k^{S_1}) = \sum_{a=1}^{F_c} P(A_a^{X_1} \mid X_1) p(\lambda_g \frac{a}{W}, a-k). \tag{7.7}$$

As stated above, the terms $P(A_a^{X_1} \mid X_1)$ express the probability that, given the occurrence of event X_1, X_1 occurred with exactly (a) available tracks. Each of $P(A_a^{X_1} \mid X_1)$ is defined by Bayes' rule [Fell52], according to

$$P(A_a^{X_1} \mid X_1) = \frac{P(A_a^{X_1})}{\sum_{j=1}^{F_c} P(A_j^{X_1})}, \tag{7.8}$$

where $P(A_1^{X_1}), \ldots, P(A_{F_c}^{X_1})$ are given by Equation 7.4. Substituting 7.7 and 7.8 into 7.5, get

$$P(S_1 \mid X_1) = \sum_{k=1}^{F_c} P(A_k^{S_1}) = \sum_{k=1}^{F_c} \sum_{a=1}^{F_c} \frac{P(A_a^{X_1})}{\left[\sum_{j=1}^{F_c} P(A_j^{X_1})\right]} p(\lambda_g \frac{a}{W}, a-k). \tag{7.9}$$

7.4.2.2 The First S Block Event, for Any Value of F_s

Equation 7.9 assumes a specific value of S block flexibility, $F_s = 3$. This section shows how Equation 7.9 can be generalized for other values of F_s. In Equation 7.6, a one-to-one correspondence was assumed between the subscript (a) in $A_a^{X_1} \mid X_1$, on the left hand side of the equation, and the variable (a), on the right hand side. This relation holds for $F_s = 3$ but does not necessarily apply for other values of F_s. For example, if $F_s = 6$, a more appropriate variable for the right hand side of the equation is $2a$. In general, the subscript (a) should be scaled by some factor, α, and Equation 7.6 becomes

$$P(A_k^{S_1} \mid (A_a^{X_1} \mid X_1)) = p(\lambda_g \frac{\alpha a}{W}, \alpha a - k)\,. \qquad 7.10$$

Clearly, α depends on the value of F_s, but α may also depend on whether a connection passes straight through a particular S block, or turns. Define Z_1 as the event that a connection passes straight through an S block, and Z_2 as the event that it turns. Also, define α_1 and α_2 as the values of α corresponding to Z_1 and Z_2. Since $S_1 \mid X_1$ implies one of Z_1 and Z_2, then

$$P(S_1 \mid X_1) = P(Z_1) \cdot P((S_1 \mid X_1) \mid Z_1) + P(Z_2) \cdot P((S_1 \mid X_1) \mid Z_2)$$

and using Equation 7.9 and 7.10,

$$P(S_1 \mid X_1) =$$

$$P(Z_1) \cdot \sum_{k=1}^{W} \sum_{a=1}^{F_c} \frac{P(A_a^{X_1})}{\left[\sum_{j=1}^{F_c} P(A_j^{X_1})\right]} \cdot p(\lambda_g \frac{\alpha_1 a}{W}, \alpha_1 a - k) +$$

$$P(Z_2) \cdot \sum_{k=1}^{W} \sum_{a=1}^{F_c} \frac{P(A_a^{X_1})}{\left[\sum_{j=1}^{F_c} P(A_j^{X_1})\right]} \cdot p(\lambda_g \frac{\alpha_2 a}{W}, \alpha_2 a - k)\,. \qquad 7.11$$

Appropriate values for $P(Z_1)$ (note that $P(Z_2) = 1 - P(Z_1)$), α_1, and α_2 are discussed in Section 7.5. Note that the (k) summation in Equation 7.11 has an upper limit of W whereas the corresponding upper limit in Equation 7.9 is F_c. This change is required since it may be possible to connect to all W tracks in a channel for values of F_s that are greater than three.

7.4.2.3 The Remaining S Block Events

Thus far, this section has dealt specifically with the event $S_1 \mid X_1$, but the derived expressions are applicable to any of the other S block events, with two changes. First, for the mth S block event, $(S_m \mid S_{m-1} \cap \cdots \cap S_1 \cap X_1)$, all summations must reach an upper limit of W. Second, the probabilities $P(A_1^{X_1}), \ldots, P(A_{F_c}^{X_1})$ in Equation 7.11 are replaced by $P(A_1^{S_{m-1}}), \ldots, P(A_W^{S_{m-1}})$, which are defined by Equation 7.12, with $m = m-1$. Applying these changes, Equation 7.7 becomes

$$P(A_k^{S_m}) = \sum_{a=1}^{W} P(A_a^{S_{m-1}} \mid S_{m-1} \cap \cdots \cap S_1 \cap X_1) \cdot p(\lambda_g \frac{a}{W}, a-k) \qquad 7.12$$

and Equation 7.11 becomes

$$P(S_m \mid S_{m-1} \cap \cdots \cap S_1 \cap X_1) =$$

$$P(Z_1) \cdot \sum_{k=1}^{W} \sum_{a=1}^{W} \frac{P(A_a^{S_{m-1}})}{\left[\sum_{j=1}^{W} P(A_j^{S_{m-1}})\right]} \cdot p(\lambda_g \frac{\alpha_1 a}{W}, \alpha_1 a - k) +$$

$$P(Z_2) \cdot \sum_{k=1}^{W} \sum_{a=1}^{W} \frac{P(A_a^{S_{m-1}})}{\left[\sum_{j=1}^{W} P(A_j^{S_{m-1}})\right]} \cdot p(\lambda_g \frac{\alpha_2 a}{W}, \alpha_2 a - k). \qquad 7.13$$

7.4.3 The C Block to Logic Block Event

The event X_2 is depicted by Figure 7.6, which shows a set of $A^{S_n} = 4$ tracks, drawn as bold lines, that are available at a C block (this corresponds to the event $A_4^{S_n}$ in Section 7.4.2) and a set of $F_c = 5$ tracks that connect to the appropriate logic block pin for the connection. The event X_2 can then be viewed as a random process in which the logic block pin can be connected to any of the set of (A^{S_n}) tracks where there are switches. Stated differently, given that one or more tracks were available at the outgoing side of the last S block, it is necessary to determine the probability that one or more of these tracks connects to the appropriate logic block pin. To simplify the notation, the expression $S_n \cap \cdots S_1 \cap X_1$ will be substituted for by SX. To calculate the probability of $X_2 \mid SX$, define the opposite event $NONE \mid SX$, where $P(X_2 \mid SX) = 1 - P(NONE \mid SX)$. To find $P(NONE \mid SX)$, assume a specific value of $A^{S_n} = a$ and define the corresponding event $A_a^{S_n}$. A conditional probability for $NONE \mid SX$ can then be defined by

$$P((NONE \mid SX) \mid A_a^{S_n}) = \frac{{}_{(W-F_c)}C_a}{{}_W C_a}. \qquad 7.14$$

where ${}_W C_a$ means the combinations of (W) things taken (a) at a time. In words, Equation 7.14 expresses the ratio of the number of ways in which the all (a) of the available tracks can fall on the $W - F_c$ tracks that cannot connect to the logic block pin to the total number of ways in which the (a) available tracks can lie on any of the W tracks.

Equation 7.14 assumes that each of the F_c switches for the logic block pin associated with event X_2 is equally likely to be on any of the W tracks.

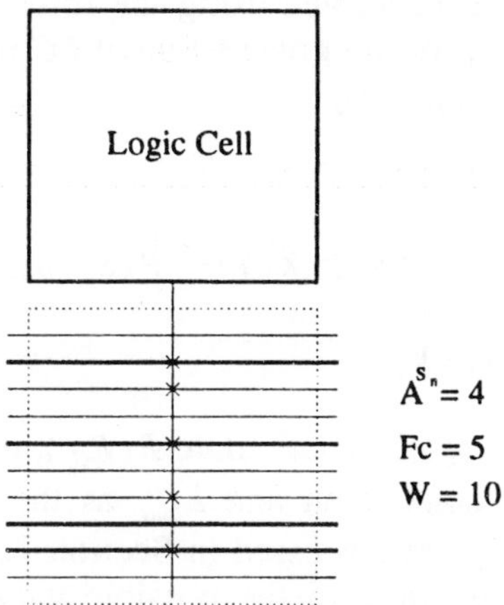

Figure 7.6 - *The Event* X_2.

This may not be realistic since a good C block topology would ensure that the tracks that are connectable to one pin would overlap the tracks connectable to others, as was discussed in Section 6.1.2.1. This assumption will have the effect of producing low predictions of routability for low values of F_c, which is discussed further in Section 7.5.

Consider the events $A_1^{S_n}, A_2^{S_n}, ..., A_W^{S_n}$ corresponding to the possible values of A^{S_n}. Since the occurrence of $NONE \mid SX$ implies exactly one of $A_1^{S_n}, A_2^{S_n}, ..., A_W^{S_n}$, it follows that

$$P(NONE \mid SX) = \sum_{a=1}^{W} P(A_a^{S_n} \mid SX) \cdot P((NONE \mid SX) \mid A_a^{S_n}),$$

where each of $P(A_a^{S_n} \mid SX)$ is given by Bayes' rule, so that

$$P(X_2 \mid SX) = 1 - P(NONE \mid SX) = 1 - \sum_{a=1}^{W} \frac{P(A_a^{S_n})}{\left[\sum_{j=1}^{W} P(A_j^{S_n})\right]} \cdot \frac{{}_{(W-F_c)}C_a}{{}_{W}C_a}. \quad 7.15$$

Each of $P(A_1^{S_n}), ..., P(A_W^{S_n})$ can be calculated using Equation 7.12, with $m = n$. Note that for the case of a connection that has length one, there are no S block events, so that $A_a^{S_n}$ in Equation 7.15 are replaced by $A_a^{X_1}$. Each of $P(A_1^{X_1}), ..., P(A_{F_c}^{X_1})$ can be calculated using Equation 7.4.

7.4.4 The Probability of R_{C_i}

Equation 7.1 can now be solved using the formulas developed in this section to calculate $P(R_{C_i})$, for the given value of $LC_i = n+1$. Equation 7.1 is reproduced below, as Equation 7.16.

$$P(R_{C_i} \mid L_{n+1}) = P(X_1 \cap S_1 \cap S_2 \cdots \cap S_n \cap X_2)$$

$$= P(X_1)P(S_1 \mid X_1)P(S_2 \mid S_1 \cap X_1) \cdots P(S_n \mid S_{n-1} \cap \cdots \cap S_1 \cap X_1)$$

$$\cdot P(X_2 \mid S_n \cap \cdots S_1 \cap X_1). \qquad 7.16$$

To make use of this result to calculate $P(R_{C_i})$, define $LC_i = l_{max}$ as the maximum length of any connection and $L_{l_{max}}$ as the corresponding event. Appropriate values for l_{max} are discussed in Section 7.5. Next, consider the events $L_1, ..., L_{l_{max}}$ corresponding to the possible values of LC_i. Since the occurrence of R_{C_i} implies exactly one of $L_1, ..., L_{l_{max}}$, then

$$P(R_{C_i}) = \sum_{l=0}^{l_{max}} P(L_l) \cdot P(R_{C_i} \mid L_l), \qquad 7.17$$

where $P(L_l)$ are given by the probability distribution of connection length, referred to in Section 7.2 as P_L, and each $P(R_{C_i} \mid L_l)$ is defined by Equation 7.16. As mentioned in Section 7.2, P_L is assumed to be geometric, with mean $\bar{R}$. Thus, $P(L_l)$ is given by

$$P(L_l) = pq^{l-1},$$

where $p = \frac{1}{\bar{R}}$ and $q = 1-p$. The following section shows how Equation 7.17 is evaluated to predict routability.

7.5 Using the Stochastic Model to Predict Routability

In order to make use of Equation 7.17, it is necessary to choose appropriate values for the various parameters that appear in the formulas developed in Section 7.4, as well as to evaluate the function λ_g, that is used to predict channel densities. This section first shows how λ_g is calculated and then gives appropriate values for each of the parameters. Following this, the routability predictions that are produced by the stochastic model are presented, with comparisons to the experimental results that were shown in Chapter 6.

As stated in Section 7.3, the parameter λ_g is defined by $\lambda_g = \frac{\lambda \bar{R}}{2}$, where

$\bar{R}$ is the average connection length and λ is the ratio of the expected number of routed connections to the total number of logic blocks. Given this definition, λ must be re-calculated after each connection is probabilistically 'routed' by the stochastic process. Thus, after $i-1$ connections have been 'routed', λ can be calculated as

$$\lambda = \frac{1}{N^2} \sum_{c=1}^{i-1} P(R_{C_c}) . \tag{7.18}$$

It is necessary to assign values to the following parameters: N W l_{max}, C_T, $\bar{R}$, $P(Z_1)$, α_1, α_2, F_s, and F_c. The first three of these depend on the size of the FPGA array and the next three are determined by the characteristics of the circuit to be routed. Since the routability predictions that are generated in this chapter are to be compared with the results from Chapter 6, the parameters will be taken from the FPGA circuits that were used in that chapter. The corresponding values are given in Table 7.1.

Circuit	N	W	l_{max}	C_T	$\bar{R}$	$P(Z_1)$
BUSC	11	11	20	392	2.7	.71
DMA	15	12	28	771	2.8	.75
BNRE	20	14	38	1257	3.0	.75
DFSM	21	13	40	1422	2.85	.76
Z03	25	13	48	2135	3.15	.75

Table 7.1 - *Stochastic Model Parameters for Experimental Circuits.*

The parameters α_1 and α_2 can be approximated by making some assumptions concerning the topology of the S blocks. It is assumed here that the topology is similar to the one used in Chapter 6. This means that, as F_s is increased from its minimum value of 2, switches are added to the wire segments in the order straight across, right turn, left turn, straight across, right turn, etc.. It is further assumed that the topology spreads the switches among

	F_s									
	2	3	4	5	6	7	8	9	10	...
α_1	1.0	1.0	2.0	2.0	2.0	3.0	3.0	3.0	4.0	...
α_2	0.5	1.0	1.0	1.5	2.0	2.0	2.5	3.0	3.0	...

Table 7.2 - *Approximations to α_1 and α_2.*

the tracks such that every track can be switched to exactly F_s others. Given these assumptions, appropriate values of α_1 and α_2 are shown in Table 7.2.

7.5.1 Routability Predictions

Recall, from Section 7.2, that routability is defined as

$$Routability = \frac{1}{C_T} \sum_{i=1}^{C_T} P(R_{C_i}), \qquad 7.19$$

This equation can now be evaluated using Equation 7.17, the formulas developed in Section 7.4, Equation 7.18, and Tables 7.1 and 7.2. A typical result is shown in Figure 7.7, which gives a plot of the expected percentage of successfully completed connections versus connection block flexibility, F_c, for parameters that correspond to the circuit BNRE.

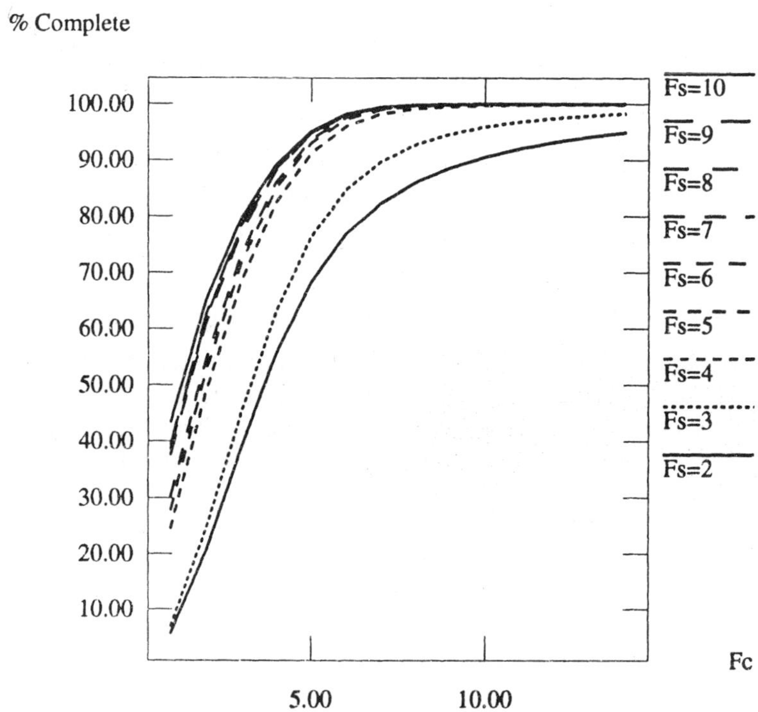

Figure 7.7 - *Routability Predictions vs. F_s, for Circuit BNRE.*

Figure 7.7 is analogous to Figure 6.9. Each curve in the figure corresponds to a different value of S block flexibility, F_s. The lowest curve represents the case $F_s = 2$ and the highest curve corresponds to $F_s = 10$. The figure indicates that the routability is low for small values of F_c and only approaches 100% when F_c is at least one-half of W. The figure also shows that increasing the S block flexibility improves the completion rate at a given F_c, but to get near 100% the value of F_c must always be high (above 7 for this circuit). The reader will note that these are the same conclusions that were reached experimentally in Chapter 6.

Figure 7.8 is a plot of the expected percentage of successfully completed connections versus S block flexibility, F_s, also for the circuit BNRE. This figure is analogous to Figure 6.12. Each curve in the figure corresponds to a different value of F_c, with the lowest curve representing $F_c = 1$ and the highest curve corresponding to $F_c = W$. The curves show an increase in slope at F_s values of 4, 7, and 10. This occurs because switches are added straight

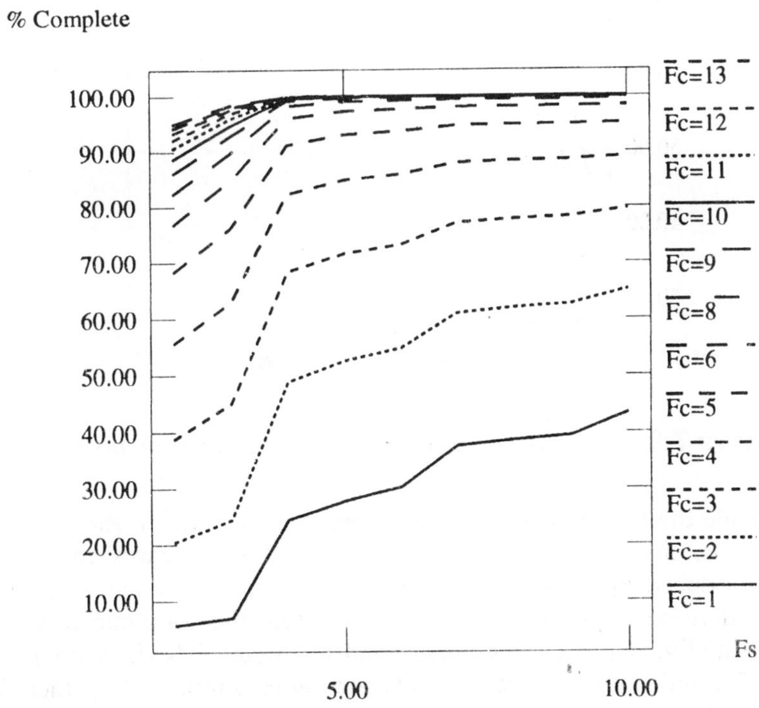

Figure 7.8 - *Routability Predictions vs. F_c, for Circuit BNRE.*

across the S blocks for these values of F_s and, as Table 7.1 shows, connections pass straight through the S blocks more than 70 percent of the time. Figure 7.8 shows that if F_c is at least half of W then very low values of F_s approach 100% routability. Again, this is the same conclusion reached in Chapter 6.

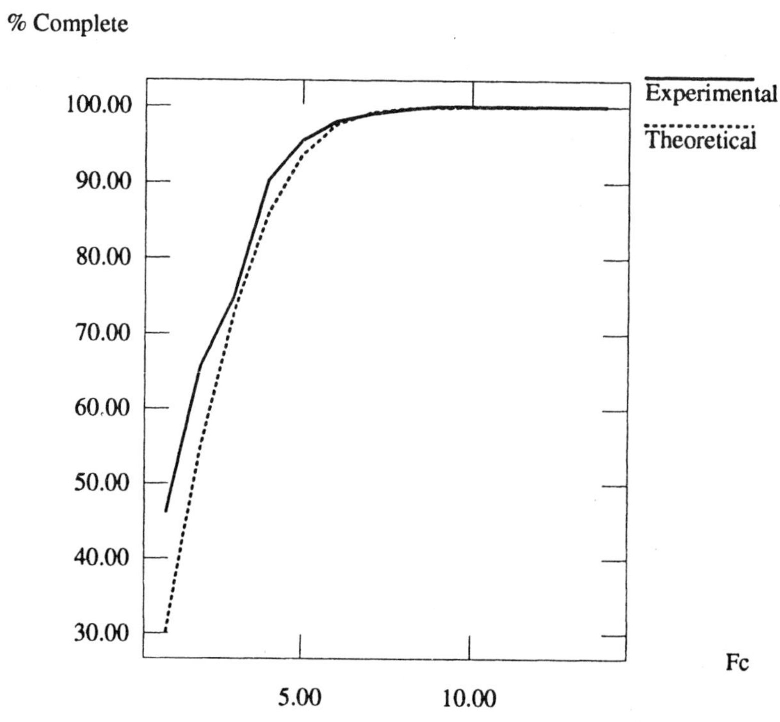

Figure 7.9 - *Comparison of Predictions and Experiments for* $F_s = 6$.

While the theoretical and experimental results lead to the same general conclusions, they are not identical. Figure 7.9 compares the routability results produced by the stochastic model with the experimental results. The dashed curve corresponds to the model, whereas the solid curve is produced experimentally. Both curves correspond to circuit BNRE, with $F_s = 6$. As Figure 7.9 indicates, the two results are quite similar. The fact that the theoretical curve is lower than the experimental curve for low values of F_c is due in part to Equation 7.14, which, as discussed in Section 7.4, does not accurately represent good C block topologies.

A summary of comparisons between theory and experiment for all of the circuits appears in Table 7.3. For each circuit, the table shows the difference between the theoretical and experimental routability results, for each value of F_s. Each entry gives the mean value and standard deviation of the difference between the experimental and theoretical routabilities, over the range of values of F_c from 1 to W. The values in the table are in percentages since those are the units of routability. Absolute values are used in the table to avoid a misleading average that could be caused by combining negative and positive differences. However, this is not really necessary since, as Figure 7.9 indicates, the theoretical predictions are almost always pessimistic. As Table 7.3 shows, the experimental measurements and theoretical predictions of routability are surprisingly close, especially for values of F_s greater than three.

	BUSC		DMA		BNRE		DFSM		Z03	
F_s	Mean	S.D.	Mean	S.D	Mean	S.D.	Mean	S.D.	Mean	S.D.
2	7.7	4.9	10.2	7.5	7.3	6.3	8.9	8.3	7.2	5.9
3	9.7	5.6	12.5	8.1	8.7	6.7	10.8	9.4	10.2	5.6
4	2.9	2.9	4.1	4.5	1.5	3.1	2.7	5.3	1.9	2.1
5	3.7	4.3	4.9	5.8	2.4	4.3	3.7	6.1	1.8	2.7
6	3.2	3.5	5.0	6.2	2.6	4.7	4.0	6.7	2.1	3.3
7	4.8	4.3	5.1	6.6	2.8	4.3	3.9	6.1	1.8	2.8
8	4.3	4.6	5.1	6.5	3.1	4.3	4.1	6.2	2.2	2.6
9	4.3	4.9	5.0	6.3	3.2	4.4	4.2	6.2	2.5	3.0
10	4.3	4.8	5.2	6.7	3.2	4.3	4.2	5.9	2.9	3.4

Table 7.3 - *Summary of Comparisons Between Theory and Experiment.*

7.6 Final Remarks

This chapter has described a theoretical approach that can be used to study FPGA routing architectures. The stochastic model presented assumes a symmetrical FPGA whose tracks consist of short segments that span the size of one logic block. It has been shown that the model produces predictions of routability that are close to those generated by the experiments shown in Chapter 6. In future work, the model could be extended so that it could also handle tracks with longer segmentations allowing theoretical studies of such architectures. This would require a slightly different approach for calculating the probability of successfully routing each connection, and to the method used for approximating channel densities.

Extending the stochastic model so as to allow more general types of routing architectures would be a worthwhile effort since this would facilitate a broader study of FPGA routing architectures. This is important because FPGAs are still very new and much research is required before the best designs will be discovered.

References

[Abou90]

P. Abouzeid, L. Bouchet, K. Sakouti, G. Saucier and P. Sicard, ''Lexicographical Expression of Boolean Function for Multilevel Synthesis of high Speed Circuits,'' *Proc. SASHIMI 90*, Oct. 1990, pp. 31-39.

[Aho85]

A. Aho, M. Ganapathi, ''Efficient tree pattern matching: an aid to code generation,'' *12th ACM Symposium on Principles of Programming Languages*, Jan. 1985, pp.334-340.

[Ahre90]

M. Ahrens, A. El Gamal, D. Galbraith, J. Greene, S. Kaptanoglu, K. Dharmarajan, L. Hutchings, S. Ku, P. McGibney, J. McGowan, A. Samie, K. Shaw, N. Stiawalt, T. Whitney, T. Wong, W. Wong and B. Wu, ''An FPGA Family Optimized for High Densities and Reduced Routing Delay,'' *Proc. 1990 Custom Integrated Circuits Conference*, May 1990, pp. 31.5.1 - 31.5.4.

[Aker72]

S.B. Akers, ''Routing,'' Chapter 6 of *Design Automation of Digital Systems; Theory and Techniques*, M.A. Breuer, Ed., NJ, Prentice-Hall, 1972.

[Alt90]

The Maximalist Handbook, Altera Corp., 1990.

[AMD90]

MACH 1 and MACH 2 Device Families Preliminary Data Sheets, 1990.

[Beda92]

A. Bedarida, S. Ercolani, G. De Micheli, "A New Technology Mapping Algorithm for the Design and Evaluation of Fuse/Antifuse-based Field-Programmable Gate Arrays," in FPGA '92, *ACM/SIGDA First International Workshop on Field-Programmable Gate Arrays*, Berkeley, CA, pp. 103-108.

[Berg88]

R.A. Bergamaschi, "Automatic Synthesis and Technology Mapping of Combinational Logic," *Proc. ICCAD 88*, Nov 1988, pp.466-469.

[Berk88]

M. Berkelaar and J. Jess, "Technology Mapping for Standard Cell Generators," *Proc. ICCAD 88*, Nov 1988, pp. 470-473.

[Bert92]

P. Bertin, D. Roncin, J. Vuillemin, "Programmable Active Memories: Performance Measurements," in FPGA '92, *ACM/SIGDA First International Workshop on Field-Programmable Gate Arrays*, Berkeley, CA, February 1992, pp. 57-59.

[Birk91]

J. Birkner, A. Chan, H.T. Chua, A Chao, K Gordon, B. Kleinman, P. Kolze, R. Wong, "A Very High-Speed Field Programmable Gate Array Using Metal-to-Metal Anti-Fuse Programmable Elements," New Hardware Product Introduction at CICC '91 *Custom Integrated Circuits Conference 91*, May 1991.

[Bost87]

D. Bostick, G. D. Hachtel, R. Jacoby, M. R. Lightner, P. Moceyunas, C. R. Morrison, D. Ravenscroft, "The Boulder Optimal Logic Design System," *Proc. ICCAD-87*, Nov. 1987, pp. 62-65.

[Bray82]

R. K. Brayton and C. McMullen, "The Decomposition and Factorization of Boolean Expressions," *Proc. International Symposium on Circuits and Systems*, May 1982, pp. 49-54

[Bray84]

R.K. Brayton et. al, **Logic Minimization Algorithms for VLSI Synthesis**, Kluwer Academic Publishers, 1984.

[Bray86]

R. Brayton, E. Detjens, S. Krishna, T. Ma, P. McGeer, L. Pei, N. Phillips, R. Rudell, R. Segal, A. Wang, R. Yung and A. Sangiovanni-Vincentelli, ''Multiple-Level Logic Optimization System,'' *Proc. IEEE International Conference on Computer Aided Design*, pp. 356-359, Nov. 1986.

[Bray87]

R. K. Brayton, R. Rudell, A. Sangiovanni-Vincentelli and A. Wang, ''MIS: a Multiple-Level Logic Optimization System,'' *IEEE Transactions on CAD*, Vol CAD-6, No. 6, Nov. 1987, pp. 1062-1081.

[Breu77]

M.A. Breuer, ''Min-Cut Placement,'' *Journal of Design Automation and Fault Tolerant Computing*, pp. 343-362, Oct. 1977.

[Brow90]

S. Brown, J. Rose and Z.G. Vranesic, ''A Detailed Router for Field-Programmable Gate Arrays'', *Proc. IEEE International Conference on Computer Aided Design*, pp. 382-385, Nov. 1990.

[Brow91]

S. Brown, J. Rose and Z.G. Vranesic, ''A Detailed Router for Field-Programmable Gate Arrays '', to appear in *IEEE Transactions on Computer Aided Design of Integrated Circuits and Systems*, 1992.

[Brow92]

S. Brown, ''Routing Algorithms and Architectures for Field-Programmable Gate Arrays,'' Doctoral Dissertation, University of Toronto, January 1992.

[Brya86]

R. E. Bryant, ''Graph based algorithms for Boolean function manipulation,'' *IEEE Trans. on Computers,* C-35(8) Aug. 1986, pp. 667-691.

[Cart86]

W. Carter, K. Duong, R. H. Freeman, H. Hsieh, J. Y. Ja, J. E. Mahoney, L. T. Ngo and S. L. Sze, ''A User Programmable Reconfigurable Gate Array,'' *Proc. 1986 Custom Integrated Circuits Conference*, May 1986, pp. 233-235.

[Chow91]

P. Chow,S.O. Seo, D. Au, B. Fallah, C. Li, J.Rose, "A 1.2um CMOS FPGA Using Cascaded Logic Blocks and Segmented Routing," International Workshop on Field Programmable Logic and Applications, Sept 1991, Oxford, UK, also available as **FPGAs** W. Moore and W. Luk Eds., Abingdon EE&CS Books, 1991, pp. 91-102.

[Cong88]

J. Cong and B. Preas, "A New Algorithm for Standard Cell Global Routing," *Proc. IEEE International Conference on Computer Aided Design*, pp. 176-179, Nov. 1988.

[Detj87]

E. Detjens, G. Gannot, R. Rudell, A. Sangiovanni-Vincentelli and A. Wang, "Technology Mapping in MIS," *Proc. ICCAD 87*, Nov 1987, pp. 116-119.

[Diet80]

D.L. Dietmeyer and M.H. Doshi, "Automated PLA Synthesis of the Combinational Logic of a DDL Description," *Design Automation and Fault-Tolerant Computing*, Vol III, No. 3/4, 1980.

[Ebel91]

C. Ebeling, G. Borriello, S. Hauck, D. Song, and E. Walkup, "Triptych: A New FPGA Architecture," International Workshop on Field Programmable Logic and Applications, Sept 1991, Oxford, UK, also available as **FPGAs** W. Moore and W. Luk Eds., Abingdon EE&CS Books, 1991, pp. 75-90.

[ElAy88]

K. El-Ayat, A. El Gamal and A. Mohsen, "A CMOS Electrically Configurable Gate Array," *Int'l Solid State Circuits Conf. Digest of Technical Papers*, Feb. 1988.

[ElGa81]

A. El Gamal, "Two-Dimensional Stochastic Model for Interconnections in Master Slice Integrated Circuits" , *IEEE Transactions on Computer Aided Design of Integrated Circuits and Systems*, Vol. CAS-28, No. 2, February 1981.

[ElGa88]

A. El Gamal, J. Greene, J. Reyneri, E. Rogoyski, K. El-Ayat and A. Mohsen, "An Architecture for Electrically Configurable Gate Arrays," *Proc. 1988 Custom Integrated Circuits Conference*, May 1988, pp.

15.4.1 - 15.4.4.

[ElGa89a]

A. El Gamal, J. Kouloheris, D. How and M. Morf, "BiNMOS: A Basic Cell for BiCMOS Sea-of-Gates," *Proc. 1989 CICC*, May 1989, pp. 8.3.1-8.3.4.

[ElGa89b]

A. El Gamal, J. Greene, J. Reyneri, E. Rogoyski, K. El-Ayat and A. Mohsen, "An Architecture for Electrically Configurable Gate Arrays," *IEEE Journal of Solid State Circuits* Vol. 24, No. 2, April 1988, pp. 394-398.

[Erco91]

S. Ercolani and G. De Micheli, "Technology Mapping for Electrically Programmable Gate Arrays," *Proc. 28th DAC*, June 1991, pp. 234-239.

[Fell52]

W. Feller, **Introduction to Probability Theory and its Applications**, John Wiley and Sons, 1952.

[Filo91]

D. Filo, J. C. Yang, F. Mailhot and G. De Micheli, "Technology Mapping for a Two-Output RAM-based field Programmable Gate Array," *Proc. EDAC 91*, Feb, 1991, pp. 534-538.

[Fran90]

R.J. Francis, J. Rose and K. Chung, "Chortle: A Technology Mapping Program for Lookup Table-Based Field-Programmable Gate Arrays," *Proc. 27th Design Automation Conference*, June 1990, pp. 613-619.

[Fran91a]

R. J. Francis, J. Rose and Z. Vranesic, "Chortle-crf: Fast Technology Mapping for Lookup Table-Based FPGAs," Proc. 28th DAC, June 1991 pp. 227-233.

[Fran91b]

R. J. Francis, J, Rose and Z. Vranesic, "Technology Mapping of Lookup Table-Based FPGAs for Performance," *Proc. ICCAD-91*, Nov, 1991.

[Fran92]

R. J. Francis, Doctoral Thesis, University of Toronto, 1992.

[Gare79]

M. R. Garey, D. S. Johnson **Computers and Intractability: A Guide to the Theory of NP-Completeness,** W. H. Freeman and Co., New York, 1979.

[Green90]

J. Greene, V. Roychowdhury, S. Kaptanoglu, and A. El Gamal, "Segmented Channel Routing," *Proc. 27th Design Automation Conference*, pp. 567-572, June 1990.

[Greg86]

D. Gregory, K. Bartlett, A. de Geus and G. Hachtel, "Socrates: a system for automatically synthesizing and optimizing combinational logic," *Proc. 23rd Design Automation Conference*, June 1986, pp. 79-85.

[Gupt90]

A. Gupta, V. Aggarwal, R. Patel, P. Chalasani, D. Chu, P. Seeni, P. Liu, J. Wu and G. Kaat, "A User Configurable Gate Array Using CMOS-EPROM Technology," *Proc. 1990 Custom Integrated Circuits Conference*, May 1990, pp. 31.7.1 - 31.7.4.

[Hamd88]

E. Hamdy, J. McCollum, S. Chen, S. Chiang, S. Eltoukhy, J. Chang, T. Speers and A. Mohsen, "Dielectric Based Antifuse for Logic and Memory ICs," *International Electron Devices Meeting Technical Digest*, 1988, pp. 786-789.

[Hana72]

M. Hanan and J.M. Kurtzberg, "Placement Techniques," Chapter 4 of *Design Automation of Digital Systems; Theory and Techniques*, M.A. Breuer, Ed., NJ, Prentice-Hall, 1972.

[Hash71]

A. Hashimoto and J. Stevens, "Wire routing by optimizing channel assignment within large apertures," *Proc. 8th Design Automation Conference*, June 1971, pp. 155-163.

[Heller84]

W.R. Heller, C.G. Hsi and W.F. Mikhaill, "Wirability - Designing Wiring Space for Chips and Chip Packages," *IEEE Design and Test of Computers*, August 1984.

[Hill91]

D. Hill and N-S Woo, "The Benefits of Flexibility in Look-up Table FPGAs," in **FPGAs**, W. Moore and W. Luk Eds., Abingdon 1991, edited from the *Oxford 1991 International Workshop on Field Programmable Logic and Applications*, pp. 127-136.

[Hsie88]

H. Hsieh, K. Duong, J. Ja, R. Kanazawa, L. Ngo, L. Tinkey, W. Carter and R. Freeman, "A Second Generation User-Programmable Gate Array," *Proc. 1987 Custom Integrated Circuits Conference*, May 1987, pp. 515 - 521.

[Hsie90]

H. Hsieh, W. Carter, J. Ja, E. Cheung, S. Schreifels, C. Erickson, P. Freidin, L. Tinkey and R. Kanazawa, "Third-Generation Architecture Boosts Speed and Density of Field-Programmable Gate Arrays" *Proc. 1990 Custom Integrated Circuits Conference*, May 1990, pp. 31.2.1 - 31.2.7.

[Kahr86]

M. Kahrs, "Matching a parts library in a silicon compiler," *Proc. IEEE International Conference on Computer Aided Design*, pp. 169-172, Nov. 1986.

[Karp91a]

K. Karplus, "Xmap: a Technology Mapper for Table-lookup Field-Programmable Gate Arrays," *Proc, 28th DAC*, June 1991, pp. 240-243.

[Karp91b]

K. Karplus, "Amap: a Technology Mapper for Selector-based Field-Programmable Gate Arrays," *Proc, 28th DAC*, June 1991, pp. 244-247.

[Kawa90]

K. Kawana, H. Keida, M. Sakamoto, K. Shibata and I. Moriyama, "An Efficient Logic Block Interconnect Architecture For User-Programmable Gate Array," *Proc. 1990 Custom Integrated Circuits Conference*, May 1990, pp. 31.3.1 - 31.3.4.

[Keut87]

K. Keutzer, "DAGON: Technology Binding and Local Optimization by DAG Matching," *Proc. 24th Design Automation Conference*, June 1987, pp. 341-347.

[Koul91]

J. Kouloheris and A. El Gamal, "FPGA Performance vs. Cell Granularity," in *Proc. of Custom Integrated Circuits Conference*, May 1991, pp. 6.2.1 - 6.2.4.

[Koul92a]

J. Kouloheris and A. El Gamal, "FPGA Area vs. Cell Granularity - Lookup Tables and PLA Cells," *First ACM Workshop on Field-Programmable Gate Arrays, FPGA '92*, Berkeley, CA, February 1992, pp. 9-14.

[Koul92b]

J. Kouloheris and A. El Gamal, "FPGA Area vs. Cell Granularity - PLA Cells," to appear in *Proc. of Custom Integrated Circuits Conference*, May 1992.

[Koul92]

J. L. Kouloheris and A. El Gamal "FPGA Area versus Cell Granularity - Lookup tables and PLA Cells," *ACM/SIGDA Workshop on FPGAs (FPGA '92)*, Feb. 1992, pp. 9-14.

[Lee61]

C. Lee, "An algorithm for path connections and its applications," *IRE Transactions on Electronic Computers*, VEC-10, pp. 346-365, Sept. 1961.

[Lee88]

K. Lee and C. Sechen, "A New Global Router for Row-Based Layout," *Proc. IEEE International Conference on Computer Aided Design*, pp. 180-183, Nov. 1988.

[Loren89]

M.J. Lorenzetti and D.S. Baeder, Chapter 5 of **Physical Design Automation of VLSI Systems**, B. Preas and M. Lorenzetti, Ed., Benjamin/Cummings, 1989.

[Mail90a]

F. Mailhot, Actel Corp., *Private Communication*, 1990.

[Mail90b]

F. Mailhot and G. de Micheli, "Technology Mapping Using Boolean Matching and Don't Care Sets," *EDAC*, 1990, pp. 212-216.

[Marp92]

D. Marple and L. Cooke, "An MPGA Compatible FPGA Architecture," *ACM/SIGDA Workshop on FPGAs (FPGA '92)*, Feb. 1992, pp. 39-44.

[Marr89]

C. Marr, "Logic Array Beats Development Time Blues," *Electronic System Design Magazine*, Nov. 1989, pp. 38-42.

[MCNC91]

S. Yang, "Logic Synthesis and Optimization Benchmarks User Guide - Version 3.0," Microelectronic Center of North Carolina, Jan. 1991.

[Murg90]

R. Murgai, Y, Nishizaki, N. Shenay, R. K. Brayton and A. Sangiovanni-Vincentelli, "Logic Synthesis for Programmable Gate Arrays," *Proc. 27th DAC*, June 1990, pp. 620-625.

[Murg91a]

R. Murgai, N. Shenoy, R.K. Brayton and A. Sangiovanni-Vincentelli, "Improved Logic Synthesis Algorithms for Table Look Up Architectures," *ICCAD*, 1991

[Murg91b]

R. Murgai, N. Shenoy and R.K. Brayton, "Performance Directed Synthesis for Table Look Up Programmable Gate Arrays," *ICCAD*, 1991

[Murg92]

R. Murgai, R. K. Brayton, A.Sangiovanni-Vincentelli, "An Improved Synthesis Algorithm for Multiplexor-based PGAs," in FPGA '92, *ACM/SIGDA First International Workshop on Field-Programmable Gate Arrays*, Berkeley, CA, pp. 97-102.

[Ples89]

Plessey Semiconductor ERA60100 Advance Information, Nov. 1989.

[Plus90]

Plus Logic FPGA2020 Preliminary Data Sheet, 1990.

[Prim57]

R. Prim, "Shortest Connecting Networks and Some Generalizations," Bell System Technical Journal, Vol. 39, pp. 1389-1401, 1957.

[Rose85]

J. Rose, Z. Vranesic and W.M. Snelgrove, ''ALTOR: An Automatic Standard Cell Layout Program,'' *Proc. Canadian Conference on VLSI*, Nov. 1985, pp. 168-173.

[Rose89]

J.S. Rose, R.J. Francis, P. Chow and D. Lewis, ''The Effect of Logic Block Complexity on Area of Programmable Gate Arrays,'' *Proc. 1989 Custom Integrated Circuits Conference*, May 1989, pp. 5.3.1-5.3.5.

[Rose90a]

J. Rose, ''Parallel Global Routing for Standard Cells,'' *IEEE Transactions on Computer Aided Design* Vol. 9, No. 10, pp. 1085-1095, Oct. 1990.

[Rose90b]

J. Rose and S. Brown, ''The Effect of Switch Box Flexibility on Routability of Field Programmable Gate Arrays,'' *Proc. 1990 Custom Integrated Circuits Conference*, pp. 27.5.1-27.5.4, May 1990.

[Rose90c]

J.S. Rose, R.J. Francis, D. Lewis and P. Chow, ''Architecture of Programmable Gate Arrays: The Effect of Logic Block Functionality on Area Efficiency,'' *IEEE Journal of Solid State Circuits*, Vol. 25, No 5, October 1990, pp. 1217-1225.

[Rose91]

J. Rose and S. Brown, ''Flexibility of Interconnection Structures in Field-Programmable Gate Arrays'', *IEEE Journal of Solid State Circuits*, Vol. 26 No. 3, pp. 277-282, March 1991.

[Roth62]

J. P. Roth and R. M. Karp, ''Minimization over Boolean Graphs,'' *IBM Journal of Research and Development*, vol. 6 no. 2, April 1962, pp. 227-238.

[Rube83]

J. Rubenstein, P. Penfield and M. Horowitz, ''Signal Delay in RC Tree Networks,'' *IEEE Transactions on Computer-Aided Design of Circuits and Systems*, Vol. 2, No. 3, July 1983.

[Tseng92]

B. Tseng, J. Rose and S. Brown, ''Using Architectural and CAD Interactions to Improve FPGA Routing Architectures,'' *ACM/SIGDA*

Workshop on FPGAs (FPGA '92), Feb. 1992, pp. 3-8.

[Sech87]

C. Sechen and K. Lee, "An Improved Simulated Annealing Algorithm for Row-Based Placement," *Proc. IEEE International Conference on Computer Aided Design*, Nov. 1987, pp. 478-481.

[Sing88]

K. J. Singh, A. R. Wang, R. K. Brayton and A. Sangiovanni-Vincentelli, "Timing Optimization of Combinational Logic," *Proc. ICCAD 88*, Nov 1988, pp.282-285.

[Sing91a]

S. Singh, J. Rose, D. Lewis, K. Chung and P. Chow, "Optimization of Field-Programmable Gate Array Logic Block Architecture for Speed," in *Custom Integrated Circuits Conference 91*, CICC '91, May 1991, pp. 6.1.1 - 6.1.6.

[Sing91b]

S. Singh, "The Effect of Logic Block Architecture on FPGA Performance." M.A.Sc. Thesis, University of Toronto, 1991.

[Sing92]

S. Singh, J. Rose, P.Chow and D. Lewis, "The Effect of Logic Block Architecture on FPGA Performance," in *IEEE Journal of Solid-State Circuits,* Vol. 27 No. 3, March 1992, pp. 281-287.

[Souk81]

J. Soukup, "Circuit Layout," *Proc. of the IEEE*, Vol. 69, No. 10, pp. 1281-1304, October 1981.

[Vlad81]

A. Vladimirescu, K. Zhang, A. Newton, D.O. Pederson and A. Sangiovanni-Vincentelli, *SPICE Version 2G, User's Guide*, Department of Electrical Engineering, Univ. of California, Berkeley, August, 1981.

[Vuil91]

Jean-Michel Vuillamy, "Performance Enhancement in Field-Programmable Gate Arrays," *M.A.Sc.* Thesis, Department of Electrical Engineering, University of Toronto, April 1991.

[Woo91a]

N. Woo, "A Heuristic Method for FPGA Technology Mapping Based on Edge Visibility," *Proc. 28th DAC*, 1991

[Woo91b]

N-S. Woo, "A Study on the Structure of the Intermediate Network in FPGA Technology Mapping," in **FPGAs**, W. Moore and W. Luk Eds., Abingdon 1991, edited from the *Oxford 1991 International Workshop on Field Programmable Logic and Applications, pp. 170-178.*

[Wong89]

S.C. Wong, H.C. So, J.H. Ou and J. Costello, "A 5000-Gate CMOS EPLD with Multiple Logic and Interconnect Arrays," *Proc. 1989 Custom Integrated Circuits Conference*, May 1989, pp. 5.8.1 - 5.8.4.

[Xili89]

The Programmable Gate Array Data Book, Xilinx Co., 1989.

Index